Cambridge Primary
Revise for Primary Checkpoint

Science

Rosemary Feasey
and Andrea Mapplebeck

Study Guide

HODDER
EDUCATION
AN HACHETTE UK COMPANY

Acknowledgements

The Publishers would like to thank the following for permission to reproduce copyright material. Every effort has been made to trace or contact all copyright holders, but if any have been inadvertently overlooked the Publishers will be pleased to make the necessary arrangements at the first opportunity.

Photo acknowledgements

p. 7 *bc* © Jane Vasileva/Adobe Stock Photo; **p. 8** *bc*, **p. 42** *br*, **p. 43** *cc* © Volod 2943/Adobe Stock Photo; **p. 35** *tr* © Lucato/Adobe Stock Photo; **p. 40** *tr* © baibaz/Adobe Stock Photo; **p. 42** *bc* © Hachette UK; **p. 48** *cl* © Dorling Kindersley/UIG/Science Photo Library; **p. 50** *bl* © Oqba/Adobe Stock Photo; **p. 50** *bc* © Dimedrol 68/Adobe Stock Photo; **p. 50** *bc* © Sirichai Puangsuwan/Adobe Stock Photo; **p. 50** *bc* © Guan Jiangchi/Adobe Stock Photo; **p. 50** *br* © S Veta/Adobe Stock Photo; **p. 51** *cr* © Smuki/Adobe Stock Photo; **p. 51** *cr* © Frances Roberts/Alamy Stock Photo; **p. 51** *br* © Maksims/Adobe Stock Photo; **p. 52** *tr* © JG Domke/Alamy Stock Photo; **p. 52** *br* © Ivan Smuk/Alamy Stock Photo; **p. 60** *cc* © Rangizzz/Adobe Stock Photo; **p. 61** *cr* © Oksana/Adobe Stock Photo; **p. 70** *br* © JJ Osuna Caballero/Alamy Stock Photo; **p. 81** *cl* © Macro Vector/Adobe Stock Photo; **p. 82** *cc* © Good Studio/Adobe Stock Photo; **p. 84** *cr* © Irant Zuarb/Adobe Stock Photo; **p. 86** *tl* © Sich Karenko_com/Adobe Stock Photo; **p. 86** *tc* © Billion Photos.com/Adobe Stock Photo; **p. 86** *cr* © Shutter Man99/Adobe Stock Photo; **p. 86** *cl* © Josianne Buttigieg/EyeEm/Adobe Stock Photo; **p. 86** *cr* © Trutta/Adobe Stock Photo; **p. 89** *cl* © Chere Zoff/Adobe Stock Photo; **p. 89** *cc* © Alex Star/Adobe Stock Photo; **p. 89** *cc* © Jamroen Jaiman/Adobe Stock Photo; **p. 89** *cc* © BillionPhotos.com/Adobe Stock Photo; **p. 89** *cr* © Ivan Burchak / Alamy Stock Vector/Alamy Stock Photo; **p. 89** *bl* © Mars58/Adobe Stock Photo; **p. 89** *bc* © S Viatlan/Adobe Stock Photo; **p. 89** *bc* © Trek and Photo/Adobe Stock Photo; **p. 89** *bc* © Ivonne Wierink/Adobe Stock Photo; **p. 90** *br* © Dneutral Han/Getty Images; **p. 99** *tl* © Iaroshenko/Adobe Stock Photo; **p. 99** *tc* © Atoss/Adobe Stock Photo; **p. 99** *cl* © Hum Images/Alamy Stock Photo; **p. 99** *cl* © S Katzenberger/Adobe Stock Photo; **p. 100** *cr* © Ammit/Adobe Stock Photo; **p. 101** *cc* © Michal 812/Adobe Stock Photo; **p. 101** *cr* © Epitavi/Adobe Stock Photo; **p. 113** *cr* © Ionescu Bogdan/Adobe Stock Photo; **p. 122** *tr* © Gudella Photo/Adobe Stock Photo; **p. 123** *cr* © NASA Image Collection/Alamy Stock Photo.

t = top, *b* = bottom, *l* = left, *r* = right, *c* = centre

This text has not been through the Cambridge International endorsement process.

Although every effort has been made to ensure that website addresses are correct at time of going to press, Hodder Education cannot be held responsible for the content of any website mentioned in this book. It is sometimes possible to find a relocated web page by typing in the address of the home page for a website in the URL window of your browser.

Hachette UK's policy is to use papers that are natural, renewable and recyclable products and made from wood grown in well-managed forests and other controlled sources. The logging and manufacturing processes are expected to conform to the environmental regulations of the country of origin.

Orders: please contact Hachette UK Distribution, Hely Hutchinson Centre, Milton Road, Didcot, Oxfordshire, OX11 7HH. Telephone: +44 (0)1235 827827. Email education@hachette.co.uk. Lines are open from 9 a.m. to 5 p.m., Monday to Saturday, with a 24-hour message answering service. You can also order through our website: www.hoddereducation.com

© Rosemary Feasey and Andrea Mapplebeck 2022

This edition published in 2022 by

Hodder Education
An Hachette UK Company
Carmelite House
50 Victoria Embankment
London EC4Y 0DZ

www.hoddereducation.com

Impression number   10 9 8 7 6 5 4 3 2 1
Year   2026 2025 2024 2023 2022

Cover illustration by Lisa Hunt, The Bright Agency

Illustrations by Natalie and Tamsin Hinrichsen, Vian Oelofsen, Stéphan Theron, James Hearne

Typeset in FS Albert 12/14 by IO Publishing CC

Printed in Spain

A catalogue record for this title is available from the British Library.

ISBN: 9781398364233

Contents

Introduction

What is this book about?

This Study Guide will help you to recall key information and ideas and build your understanding about the science topics that you have been learning in Stage 6. It will help you to make sure that what you learn stays in your memory for a long time.

How can this book help you?

Revision helps you to remember facts and to practise different ways of working. It can help you make connections between ideas so that the knowledge you have and the understanding you develop is more likely to stay in your memory for longer. Revision can help to make it easier for your brain to retrieve (get back) what you know when you need it in the future, for example, in a new topic or a school test.

How can I help myself when using this book?

Be honest, if you do not know something, admit it yourself. If you still do not understand an idea or a word, ask a partner or an adult to help you understand. Positive learners know that asking for help is a good thing to do.

It's OK to make mistakes

When we learn it is fine to make mistakes and errors, especially when revising. These are great learning opportunities and should be explored if new learning is to take place. If you are unsure, do not know or make a mistake, do not worry, look back through your books, ask a partner or an adult to help you, that way you will continue to learn and succeed.

Worked examples

Throughout this Study Guide, you will find examples of learners' work which have errors in them called worked examples. Finding and correcting the errors is a great way to help you learn. You will also find examples of 'good' answers, to help you understand how questions and problems can be answered.

Model icon
Shows you are using a mental or physical model of something in the real world.

Star icon
This tells you that you need to think and work like a scientist.

Link icon
This tells you that content is related to another subject you are learning.

What approaches (ways) does this Study Guide use to help you revise science?

This Study Guide offers different approaches that you can use to help you decide what you know and understand and what you are still unsure of, this will enable you to practise and revise, ensuring that those ideas 'stick' in your memory.

There are many different ways of revising what you know in science. To help you understand how each approach works, the following pages explain each of the different approaches. Each person learns and remembers their science in different ways, approaches that you find helpful might be different to what others prefer to use.

A few of the revision approaches are explained so that you know what they are and why they are used. You might find some of them useful in other subjects to help you remember and revise facts and ideas for example, in maths, geography and history.

Concept maps

A concept map is a type of thinking map. It is a useful way to help you remember key words and ideas in a topic and make links between them.

With a concept map you can group words using headings. You can then show how ideas link by drawing lines between them. You need to write on the lines why you are connecting the lines together.

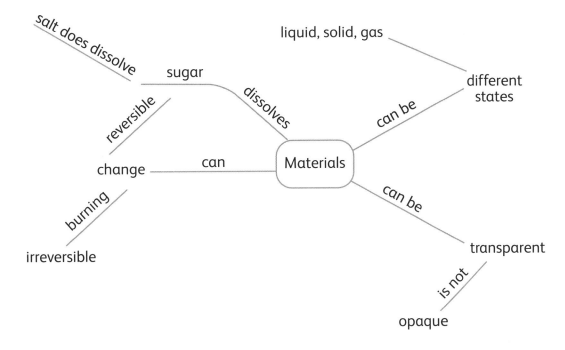

Definitions

Thinking about and writing down definitions for words can help you understand the word better and give you confidence to use the words when speaking or writing about science.

Being able to use scientific vocabulary correctly is very important, if you are unsure of a word use a dictionary or thesaurus to check if you are correct. You could make yourself a science dictionary, write the word, how to say it, a definition and even draw or stick a picture or sample of something, for example, a material (aluminium foil).

This Study Guide also provides other ways to help you learn words, such as key word cards, mnemonics and acrostics.

Double bubble

A double bubble is another type of thinking map. A double bubble is a useful way to check that you have understood ideas. The double bubble helps you look at two different ideas and compare them for similarities and differences. It is used to compare and contrast objects, animals, ideas or events.

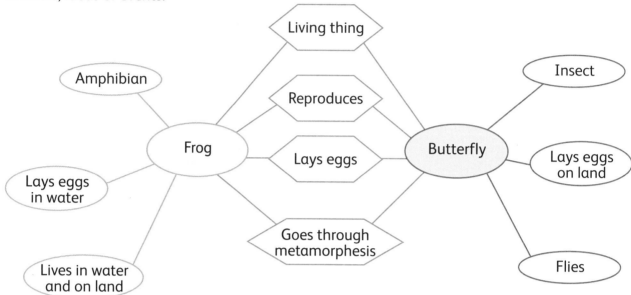

In the centre of the large circles write the name of, for example, the two ideas you are thinking about. Down the middle of the double bubble, in the hexagons, you are looking for similarities. This means you will write one thing in each hexagon that is true for both ideas. In the outside circles you then consider differences. You will note down things that are different for each of the ideas you are thinking about.

By drawing out your thinking using this type of thinking map you will make connections and remember what you have been learning about.

Frayer map

A Frayer map helps you to identify and define unfamiliar concepts and vocabulary in a topic. Learning vocabulary so that you know how to spell words and understand what they mean is very important in science. If you can read, spell and explain scientific words and apply them to everyday life, it means that your learning is secure.

A Frayer map is used to focus on revising and learning about one word at a time.

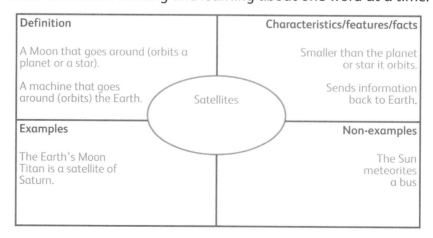

Fishbone organiser

This is another type of thinking map. A fishbone organiser is a useful way of organising ideas and knowledge linked to the same topic. It is a visual way to link many ideas that connect to the same issue and help you record what you know about a problem and how to solve it. A fishbone organiser is a simple way that you can organise your ideas, make links and solve problems. You might want to use different colours on your fishbone organiser so you can pick out ideas that link.

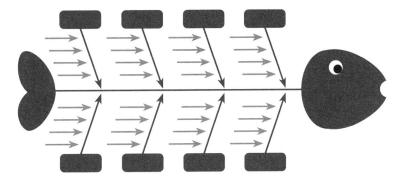

Infographics

The word 'infographic' is made up of:

info (information) + **graphic** (visual)

So, an infographic is made using pictures, charts and graphs so that information can be read easily.

Using infographics can help your brain remember information, as it draws on the idea of Dual Coding. 'Dual Coding' gives two different ways of showing the information, both visual and written at the same time. Looking at the information on an infographic and talking about it with others, can make it easier for your brain to recall (remember) information. You might find that using infographics made by others or by yourself can help you to learn ideas and facts.

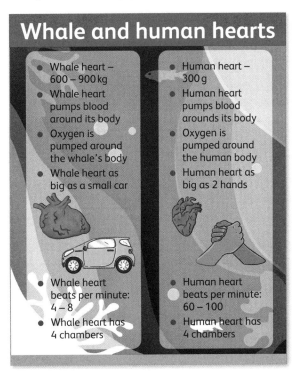

Whale and human hearts

- Whale heart – 600 – 900 kg
- Whale heart pumps blood around its body
- Oxygen is pumped around the whale's body
- Whale heart as big as a small car

- Whale heart beats per minute: 4 – 8
- Whale heart has 4 chambers

- Human heart – 300 g
- Human heart pumps blood arounds its body
- Oxygen is pumped around the human body
- Human heart as big as 2 hands

- Human heart beats per minute: 60 – 100
- Human heart has 4 chambers

Key Word Cards

Key word cards can help you remember words and revise scientific vocabulary that you need to read, spell and know what the word means. You can make key word cards for any topic, on one side write the word, on the other side write the definition (what the word means). You could write words in different colours and split words up to help you remember them. This revision approach will help you by challenging you, making sure that you can read, spell and understand what the word means.

There must be two objects that touch. These can change the motion, direction and shape of the object they act on. They can be balanced or unbalanced.	 **Contact force**
There must be two objects that interact, these objects are not touching. Examples of non-contact forces are weight (gravity) and magnetism. They can be balanced or unbalanced.	**Non-contact force** 

Fact files

Fact files are really good ways to learn, memorise and help to recall information. They are quick and easy to make so that you can show key ideas, scientific vocabulary and give examples of science in action in everyday life.

You can make a set of fact files for a topic or different topics which you can keep and use to revise learning, and to give a partner and family to use to 'test' you on your science.

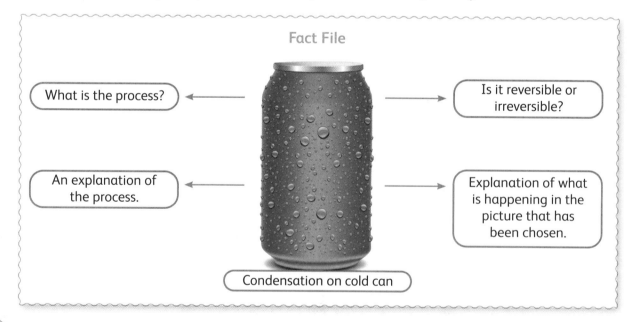

Fact File

What is the process? ← → Is it reversible or irreversible?

An explanation of the process. ← → Explanation of what is happening in the picture that has been chosen.

Condensation on cold can

Learning flower

A learning flower is a memory aid that can help your brain remember and organise ideas. This is a visual way of organising your learning, which some learners find really useful when trying to revise a topic.

This will help you recall important things you have covered in this topic and make links between the different ideas.

When you use a learning flower, write:

1 the topic title in the centre of the flower

2 key words on the front of the petals

3 definitions of the words on the back of the petals

4 key learning on the stem

5 what you already know on the roots

6 questions or things that you do not fully understand on the leaves.

Memory maps

A memory map is a way of helping you to revise information and ideas. A memory map can help you group and organise information to help your memory so that you can work out what you know already and where the gaps are in your understanding.

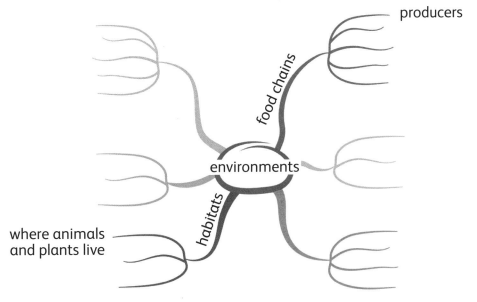

Mnemonics

Mnemonics often use rhymes or a sentence to help us learn information so that it sticks in our memory and also makes it easier to retrieve (get back) that information.

For example, you might know all of the names of the planets in our solar system, a mnemonic could help you to remember not only the name of each planet easily but the order of the planets as they orbit the Sun.

Mnemonic for the planets

My (Mercury) Very (Venus) Easy (Earth) Method (Mars) Just (Jupiter) Speeds (Saturn) Up (Uranus) Nothing (Neptune).

Model answer (Worked answer)

A model answer is an 'ideal' response to a question. Looking at different answers to questions and thinking about how they can be improved using your own knowledge is one way of helping you to revise and remember your science. Thinking about what the strengths and weaknesses of an answer are is a way of assessing how good an answer is. Re-writing it to improve the answer can help you later when you have to write an answer yourself.

Prefixes

A prefix is a group of letters added to the beginning of a word to change the meaning of the word.

For example:
The word 'microbe' begins with the root word **micro** which comes from Greek word **mikros** which means 'small'. Whenever you come across a word where the root word is micro the word has something to do with being small.

The word 'microscope' is made up of the root word 'micro' and the word 'scope'. *Micro* means 'small', and *scope* means 'see'. So the prefix changes the meaning of scope (to see) to mean 'to see very small (microscopic) things'.

How can prefixes help you in revision? If you know what some prefixes mean, for example: micro (small) or therm (heat), then it can help you to use your memory to work out what certain words mean when you read or hear them.

Rich picture poster

A rich picture is a way of showing an idea, information, a process or for example, a habitat by using pictures, diagrams and individual words, phrases and colour coding. Using a rich picture can sometimes be easier to show what you know than, for example, writing sentences or paragraphs, especially if you are someone who learns and remembers pictures more easily.

Rich pictures do not have to be created in any order, they can just show the flow of ideas, for example, how to solve a problem or how to get people to recycle and reuse materials. It is different to an infographic because it does not have to use graphs, charts and numbers.

Revision hexagons

Revision hexagons are hexagon shapes that tesselate (link together). The ideas and facts that you write in them must link to what is written in the hexagons around.

This revision approach helps you to:

- think about what you know
- recall facts and ideas
- make links between learning.

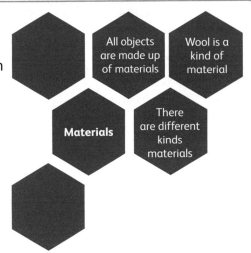

Venn diagram

A Venn diagram is a visual organiser. It helps you to see how you can organise what you know by helping to show similarities and differences between ideas.

For example, you could use a Venn diagram to organise objects according to the material/s they are made from. Venn diagrams usually have 2 or 3 circles which overlap in the middle; where the circles overlap shows similarities, and where they do not overlap shows differences.

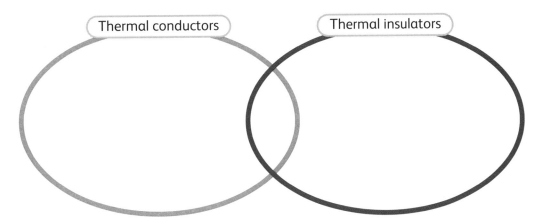

Using these different approaches in your learning and revision:

As you work through this Study Guide think about which approaches help you to:

- access your memory and remember ideas and words
- make connections (see links) between learning
- remember the most ideas or information
- organise what you know
- spot gaps in your learning and understanding.

Think about which approaches you enjoy using, if you enjoy using an approach in revision it can make revising a topic more interesting and easier to learn and remember.

Different approaches suit different learners, work out which ones suit you best, and then you can make your brain work and grow!

Systems and diseases

This unit will help you to revise your learning of:
- many vertebrates have similar circulatory systems
- how the heart works
- the function of the heart
- many vertebrates have similar respiratory systems
- how the lungs work
- the function of the lungs.

This unit will also help you to revise your learning of some diseases and their causes and defence mechanisms against infections.

 Pages 8–25

The circulatory system

Revision approach

Using memory maps

What is a memory map?

A memory map is a useful tool to revise information and ideas. In this unit, you are going to use a memory map to help you to revise information about the heart, lungs and disease, so that you can see what you know already and where the gaps are in your understanding.

Activity 1

Creating a memory map

1   In the middle of a sheet of paper or a page in your notebook, draw a picture of something that reminds you of the circulatory system. In the example below, a heart has been drawn.

You will need:
- plain paper
- coloured pens

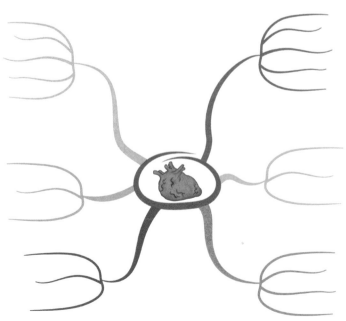

2   Draw two or three lines coming from your picture for key ideas about the circulatory system. Using different coloured pens, add as many words, phrases or pictures that you remember about the circulatory system.

3   As you continue adding to your memory map, keep going back and reading what you have written, it might help you to think of other things that you know. You can add more lines. Remember you probably know more than you think.

4   When you have finished, share your memory map with someone else. Ask them to add something that you have left out.

5   Keep your memory map, you are going to add to it throughout this unit.

6   Follow the same instructions to draw a memory map about what you can remember about the respiratory system and illness and disease.

Activity 2

1   Look at the table below.

2   Add the words from this list to the table and describe what each does in the circulatory system. One has already been done for you.

(arteries)   (blood)   (blood vessels)   (capillaries)   (lungs)   (veins)

Word	What do they do?
Arteries	Tubes that carry blood away from the heart.

3   Check your definitions using a dictionary.

4   Test yourself, or get someone else to test you, to check that you can spell each word correctly.

Activity 3

Your heart is part of the circulatory system. The diagram below shows the circulatory system.

(oxygen rich blood)   (heart)   (rest of body)   (blood low in oxygen)   (lungs)

Match the labels to the numbers.

1	
2	
3	
4	
5	

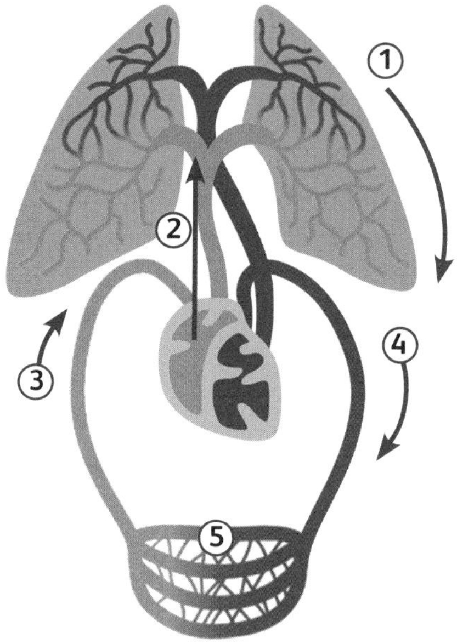

a   How many labels did you get correct? _____

b   If you got any wrong, which ones? What can you do to remember all of them correctly?

_____

_____

c   Try again, did you get them all correct this time? _____

d   Write a paragraph to explain how the circulatory system works. Use the diagram and labels to help you explain.

_____

_____

_____

_____

_____

Activity 4

Go back to your memory map of the circulatory system on page 12.
What can be added to the memory map to help you revise the circulatory system?

Activity 5

Science in context

Christiaan Barnard led the surgical team that performed the first human-to-human heart transplant on 2 and 3 December 1967. This was the first time that a human heart had been successfully transplanted into another human. It was carried out because the patient was critically ill with heart failure and would have died if he had not received a new heart from a donor.

Unfortunately, the patient who received the transplanted heart died 18 days later. However, thanks to Christiaan Barnard's pioneering work, today across the world, around 5000 cardiac (heart) transplants are carried out each year. Most of the people who have a heart transplant now live long and healthy lives, thanks to their new heart from a heart donor.

Think about what Christiaan Barnard needed to know about the heart and the circulatory system.

Write down six things that he and his team needed to know to carry out the heart transplant.

_____

_____

_____

_____

_____

_____

Activity 6

Humans and whales are both vertebrates, they both have a spine (backbone) and a circulatory system.

The poster on the next page is an **infographic** about the similarities and differences between human and whale circulatory systems. The word infographic is made up of:

(**info** (information)) + (**graphic** (visual))

So, an infographic is a visual of data and information that can be read easily.

Look at the information presented on the infographic about human and whale hearts.

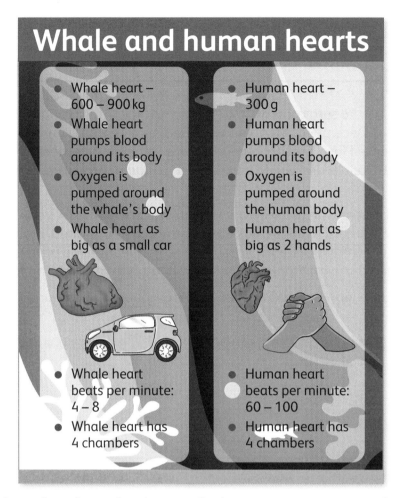

1 Add two things from the infographic that you find interesting to your circulatory system memory map.

2 Look at the table below.

3 Use the information from the infographic to complete the table.

Similarities and differences between human heart and whale heart	
Similarities	**Differences**

Activity 7

Choose three facts from the infographic that show that a human and a whale have **similar** circulatory systems and add them to your memory map.

The respiratory system

Do you remember?

The respiratory system

The function of the respiratory system is to supply every cell in the body with oxygen rich gas and remove the waste gas, carbon dioxide, which the body does not need.

The lungs are the internal organ in the respiratory system where air is taken in (inhaled) and given out (exhaled), this is called **breathing**.

When you breathe in, the diaphragm (a muscle under the lungs) moves down to make space for the lungs to fill up with the air that is breathed in through the nose and mouth (inhalation). The ribs move up and outwards to give the lungs more space. The air goes down the trachea into the lungs, to the bigger branches in the lungs (the bronchi) and then to smaller branches (bronchioles). Oxygen passes from these into the blood, and the oxygen is then carried to all the cells in the body.

The cells in the body use up the oxygen and give out carbon dioxide and other waste products. The blood going back to the heart and then to the lungs contains this carbon dioxide and waste. When you exhale (breathe out – exhalation) everything is reversed. The diaphragm moves up pushing air out of the lungs. It goes through the bronchioles, to the bronchi, up the trachea and out through the nose and mouth.

When you inhale (breathe in) air containing oxygen and exhale (send out) carbon dioxide (called expiration, or exhalation) you are exchanging oxygen and carbon dioxide.

The exchange of gases in every cell in the body is called **respiration** and the parts of the body that work together to allow this gas exchange is called the **respiratory system**. Remember, respiration is a process that is constantly repeated in order to keep us alive.

Activity 1

On a separate sheet of paper, copy and complete the flow diagram below to show respiration.

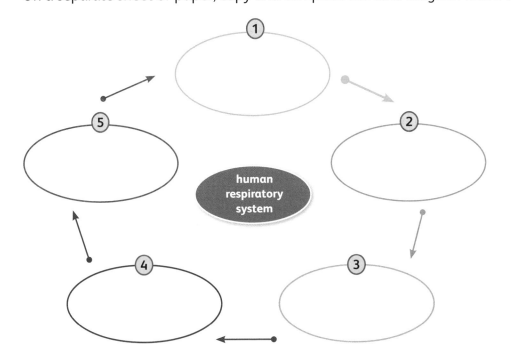

You will need:
- plain paper
- coloured pens

Activity 2

Read the information on page 17 again and list all the scientific words that explain respiration.

1   Highlight each word using a different colour.

   Red – I do not know this word.

   Orange – I know the word and can define it, but I am unsure if my definition is correct.

   Green – I know what this word means.

2   On a separate sheet of paper or in your notebook, write a definition for all the words highlighted **red** and learn how to spell them.

3   Check the definition of all the words that are **orange**, make sure you know what they mean and can spell them.

Activity 3

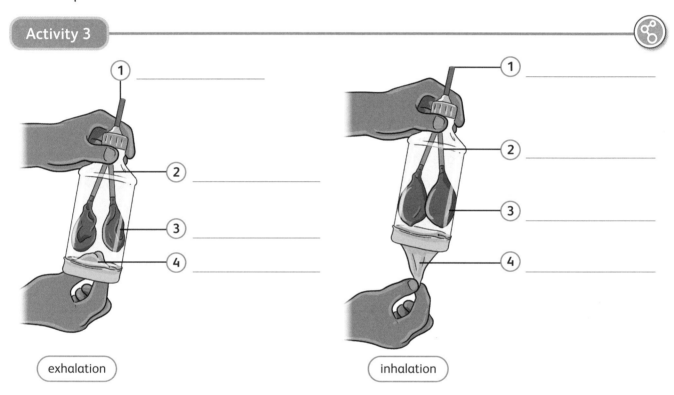

exhalation   inhalation

1   Stage 6 learners made a model to show how we breathe. Use words from the list you made in Activity 2 above to explain how the model shows inhalation and exhalation.

_____

_____

_____

2   Use the words that you listed in Activity 2 to label the parts of the model.

3   Some learners think that there are two tubes from the mouth to the lungs, instead of one tube (the trachea, also called the windpipe), which then divides into one for each lung. Look at the model above. Do you think that it would lead learners to learn an incorrect idea? Why do you think that? Explain your reason or reasons.

_____

_____

_____

_____

Activity 4

Some learners think that the words **respiration** and **breathing** mean the same thing.

1   Write a definition for each word to show the difference between respiration and breathing, so that others will understand that the words mean different things.

_____

_____

2   Look back at your respiratory system memory map. Check what you have written about the respiratory system, change anything that is incorrect and add new knowledge.

Illness and diseases

Do you remember?

Illness and diseases

There are many causes of illness and diseases, some are caused by germs which can only be seen with a microscope. These germs are called **microbes**, most microbes are harmless, but some can cause people to become ill.

Activity 1

Prefixes

A prefix is a group of letters added to the beginning of a word to change its meaning.

The word **microbe** begins with the root word **micro**, which comes from the Greek word **mikros** meaning 'small'. Whenever you come across a word where the root word is 'micro', the word has something to do with being small.

The word **microscope** is made up of the root word **micro** and the word **scope**. 'Micro' means small and 'scope' means to see. The prefix changes the meaning of 'to see' to 'to see very small (microscopic) things'.

Find four more words which use the prefix 'micro', write a definition for each one.

i   _____

_____

ii _____

_____

iii _____

_____

iv _____

_____

Activity 2

Draw lines to match these harmful microbes to the diseases. If you are unsure, do some research to find out the answer.

Harmful microbes
1   flu
2   cholera
3   athletes' foot
4   malaria

Diseases
A   bacteria
B   virus
C   parasite
D   fungus

Do you remember?

Defence mechanisms

The human body has developed defence mechanisms, these are ways the body uses to help fight infectious diseases.

How microbes can enter the body	Body's defence mechanisms
Openings such as the mouth and nose allow microbes into the stomach	Mucous in the nose captures the microbes, we then sneeze and cough to blow them out of the body; for example, when we have a cold. Acid in the stomach can kill some microbes to stop people getting an upset stomach.
Different body surfaces such as the eyes, ears	Tears flush out microbes from the eyes and wax prevents microbes entering the ears.
Breaks, cuts or insect bites on the skin	The skin itself is a barrier to infection so if it is cut or grazed, it immediately begins to heal itself, often by forming a scab.

Revision approach

Fishbone organiser

This is a fishbone organiser; it is a useful way of organising ideas and knowledge. It is a visual way to record what you know about a problem and how to solve it.

This fishbone organiser illustrates the problem of infectious diseases amongst humans. It shows the symptoms for each disease and how they are transmitted (top row), how the body's defence mechanism works against each disease and how humans can help to prevent transmission (spreading) (bottom row).

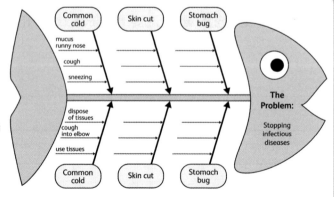

Activity 3

1 Copy the fishbone organiser or use the template provided by your teacher.

2 Complete the fishbone organiser. Begin with what you already know about each disease, including the symptoms, and write these on the top row. On the bottom row, write about the body's defence mechanisms and how humans can prevent the transmission.

You might need to carry out additional research to help you complete the fishbone organiser.

Activity 4

Go back to your memory map for this unit, change anything that you think is not correct and add facts or ideas that you think are important.

Activity 5

Below are the results of a research project to find out about hand washing and bacteria on hands. Stage 6 learners dipped their hands in dirty water and then made a handprint on special gel. Their hand-prints were left overnight and the next day each handprint showed how much bacteria had grown on the special gel. The learners guesstimated what percentage of the hand was covered in bacteria.

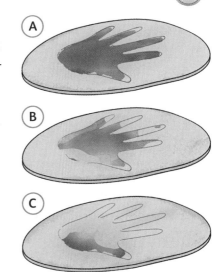

1 Look at the information in the pictures and the table.

Handwashing	A	B	C	D
	Water no soap 5 seconds	Water soap 5 seconds	Water no soap 20 seconds	Water and soap 20 seconds
	93 %	86 %	25 %	3 %

Think about what the learners were testing. What do you think the question was that they were trying to answer?

_____

_____

2 During the Covid-19 pandemic, scientists were telling people to wash their hands thoroughly for at least 20 seconds. Does the evidence from this investigation support what scientists were telling people? Why?

_____

_____

Activity 6

The diagram on the next page shows a useful way of planning a fair test.

1 On a separate piece of paper or page of your notebook, use the diagram to plan how the learners set up the test to answer the question from Activity 5.

2 When you have finished show your plan to a partner.

3   What do they think of your plan?

4   What do they think could be improved?

a   What to keep the same to make your test fair:

e   Outline of graph:

b   Prediction using scientific knowledge:

d   Table of results:

c   Relevant observations to make:

Revision quiz

Revisit the work that you have done in this unit. Read through your memory maps to help you remember information and ideas. When you are ready, complete this short quiz.

As you work through it, you can help yourself by:
- Reading each question carefully – check you understand the question.
- Look for key words, use them in your answer.
- Answer the question in your mind first before you write it.
- If the question is multiple choice, decide which answers you know are definitely incorrect, then think about the answers that are left, to decide which one is correct.
- Check your answer in case you need to change anything.

Answer the questions on a separate sheet of paper or in your notebook.

1   Circulation is the process:

   a   where blood is pumped around the body

   b   where the brain sends signals to the body

   c   where the stomach digests food

2   Respiration is:

   a   where oxygen turns into a gas

   b   where organs work together to exchange gases in every cell in the body

   c   where hairs in the nose clean the air

3   True or false?

   a   Bronchioles are found in the lungs.

   b   Exhalation is when you breathe in.

4   Match the words to the correct sentences.

   a   Circulatory system   A muscle that moves up and down as you breathe.

   b   Diaphragm   A tiny life form that can only be seen with a microscope (for example, bacteria).

   c   Microbe   Moves blood around the body.

5   What does it mean to 'transmit a disease'?

6   Give three ways that the common cold virus can be transmitted from one person to another and two ways that the body has defences to protect it.

7   Name three similarities between a whale heart and a human heart.

8   Which root word means so small that it cannot be seen by the human eye?

9   Below are four words that answer four different questions. For each word answer, write what the question could be.

   a   gas exchange

   b   respiratory system

   c   breathing

   d   respiration

10 Stage 6 learners collected data on handwashing and created this graph.

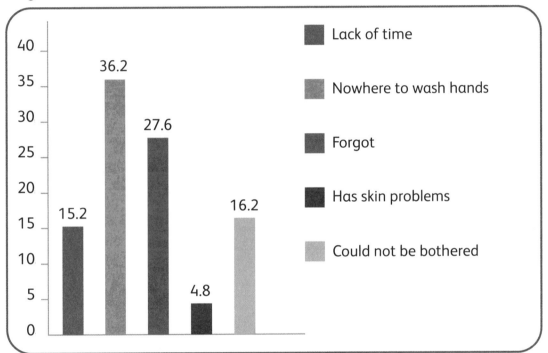

   a   What do you think their question was?

   b   Which was the most frequent answer, how do you know?

   c   List three ways that you could convince people to wash their hands regularly.

   d   What could you do to make sure that people do not forget to wash their hands?

   e   Explain why hand washing is important in preventing the transmission of diseases.

11 Explain the differences between **the circulatory system** and **the respiratory system**.

12 Explain the similarities between **the circulatory system** and **the respiratory system**.

Unit 2 Human reproduction

What will you learn

This unit will help you to revise your learning of:
- the parts of the human reproductive system
- physical changes that take place during puberty in humans.

 Pages 26–30

Revision approach

Use key word cards

What are key word cards?

Key word cards help you remember words.
A set of key word cards will help you to revise
scientific vocabulary that you need to learn to
read, spell and know the meaning of. You can
make key word cards for any topic.

How to make your own cards

When the egg
combines with the
sperm during sexual
reproduction inside
the female body.

internal fertilisation

On one side of the card write the word, on the other side write the definition (what the word
means). You could write the words in different colours or split the words to help you remember
them.

Activity 1

Below is a list of the words that you need to learn and remember about
human reproduction.

(eggs) (fallopian tubes) (gestation period) (inherits)

(internal fertilisation) (puberty) (sexual reproduction)

(sperm) (uterus)

You will need:
- small sheets of card or plain paper
- coloured pens

1 Write each word on the front of a card. You can break some words up if it helps you to
   remember how to spell them, you could even write each part in a different colour pencil or pen.

2 Draft a definition on a scrap piece of paper, or on your whiteboard. Check your definition using
   a dictionary. Think about whether your definition is a good one or if parts of it need changing,
   then write it on the back of your card.

3 Go through the key word cards, look at the word, say the definition to yourself, check the back
   of the card to see if you are correct. Now work with a partner. Ask your partner to check that
   you have memorised and know the words on your cards.

Activity 2

You will need:
- key word cards that you made in Activity 1
- 3 sticky notes or pieces of card or paper

1   On the sticky notes make three labels: *Yes, No, Unsure*.

2   Go through each of the key word cards you have made and decide if you are confident that you can spell and know the definition of each word. If you are confident, place it under the *Yes* heading, if you cannot spell a word or do not know what the word means put it under the *No* heading. If you are unsure, you think you might know how to spell the word or know the definition, place it under the *Unsure* heading.

3   Put aside the words under the *Yes* heading. You already know those words. The words under the *No* heading and the *Unsure* heading are the words that you must learn before the end of this unit. Write three ways that you could use your key word cards to learn the scientific words. For example:

- Use the word in a sentence to show that you understand its meaning.

- _____

- _____

- _____

Reproduction and puberty

Do you remember?

Reproduction

Animals, including humans, need a male and a female to reproduce (have offspring). Some animals grow their young inside their bodies and give birth to live offspring, humans do this.

Humans need a male and a female to reproduce. The male produces sperm and the female produces eggs. To create a new offspring (baby), the sperm must join with an egg, this is called **sexual reproduction**.

The female reproductive system contains tubes which are called fallopian tubes. At the end of each fallopian tube, is the ovary, the ovary contains the eggs. In males, the sperm is produced in the testicles. During sexual reproduction, the sperm travels down the penis through the vagina and cervix, through the fallopian tubes to the ovary where an egg can be fertilised by the sperm.

Because this happens inside the female body this is called **internal fertilisation**.

Once the egg has been fertilised by the sperm, the woman is pregnant, and the offspring (baby) grows inside the uterus for about 40 weeks.

Activity 1

Prefixes and suffixes

Prefixes and suffixes are helpful when learning words, especially scientific words.

A prefix is a letter or group of letters added to the beginning of a word which changes its meaning.

A suffix is a letter or group of letters added to the end of a word and changes the word's meaning.

> The words **viviparous** and **oviparous** are used in animal reproduction.
>
> The prefix **vivi** means 'living' or 'alive'.
>
> The prefix **ovi** means 'egg'.
>
> The suffix **parous** means 'having produced offspring'.

1   Choose the correct word. (viviparous)   (oviparous)

    a   Some animals lay eggs from which their young hatch, such as birds. _____

    b   Some animals grow their young inside their bodies. _____

2   Write the names of three animals that are oviparous. Explain why you have chosen these animals.

_____

_____

_____

3   Write the names of three animals that are viviparous. Explain why you have chosen these animals.

_____

_____

_____

Activity 2

Read the information in the **Do you remember?** panel on page 25. Use the information to correctly label the diagram by matching the words to the correct label number.

penis
vagina
testicles
ovary
fallopian tubes
uterus
cervix

1
2
3
4
5
6
7

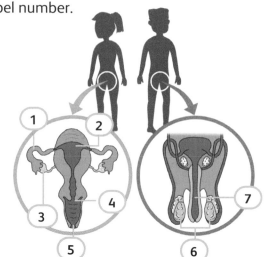

Activity 3

1 Write a definition for 'gestation period'.

2 Write a definition for the word 'inherit'.

3 Check that your definitions are correct by using your key word cards.

Activity 4

The table below shows the gestation period of different animals.

Animal	Gestation period – days (approximately)	Mass of animal kg (approximately)
Camel	360	475
Cat	60	4
Cow	280	285
Elephant	645	6000
Gerbil	22	0.058
Human	266	80
Mouse	19	0.020
Polar bear	241	1500
Rat	21	0.300
Rhino	450	2200
Sheep	144	140
Orca	510	136 000

Hint

This column shows measurement in kilograms (kg). What do you notice about the mass of the gerbil and mouse? Is it greater or less than 1 kg? How do you know?

1 Look at the data in the table.
2 Below are some conclusions that Stage 6 learners wrote. Use the data in the table to decide which conclusions are correct and incorrect.
   a The duration of gestation periods varies between different kinds of animals.
   b All animals have the same gestation period.
   c The mass of the animal does not affect the gestation period.
3 Stage 6 learners wrote the following conclusion:
Our conclusion is that the bigger the animal, the longer their gestational period.
However, they did not use the patterns in the data. Rewrite and improve their conclusion using the data in the table. When you have finished, write why you made the changes.

Do you remember?

Puberty

Puberty is the time during adolescence when a boy or girl's body begins to change and develop as they become an adult. Puberty is the time when humans reach sexual maturity. This process usually takes place between the ages 10 and 14 for girls and 12 and 16 for boys.

When the body is ready to begin puberty; it releases special hormones (chemicals that tell the body what to do), these hormones change different parts of the body depending on whether you are a boy or a girl, some of the changes are the same for both.

Some of the changes happen internally (inside the body) and some happen externally (outside the body).

Activity 5

Use the Venn diagram template from your teacher to list the changes that will happen during puberty.

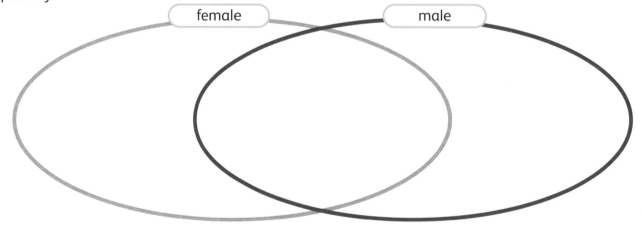

Activity 6

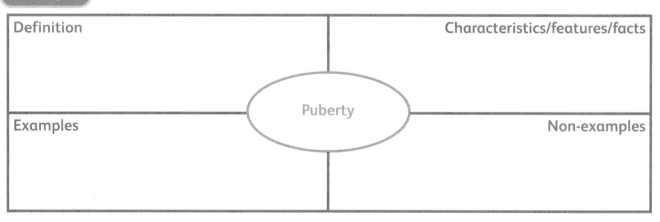

Definition	Characteristics/features/facts
Examples	Non-examples

Puberty

Look at the diagram above. This is a **Frayer map**, it helps you to organise ideas and vocabulary about a topic. Use the Frayer map template from your teacher to make your own Frayer map about puberty using the information above.

When you have completed your Frayer map, share it with a partner. They will read your map and write one thing that is good about your map and one new thing that you could add. You will do the same for their map.

Revision quiz

Revisit the work that you have done in this unit. Read through your key word cards to remember scientific words and information. Use your Frayer map to remind you about puberty. When you are ready, complete this short quiz.

As you work through it, you can help yourself by:

- Reading each question carefully – check you understand the question.
- Look for key words, use them in your answer.
- Answer the question in your mind first before you write it.
- If the question is multiple choice, decide which answers you know are definitely incorrect, then think about the answers that are left, to decide which one is correct.
- Check your answer to make sure that you do not want to make any changes.

Answer the questions on a separate sheet of paper or in your notebook.

1   Animal reproduction is the process where:
    a   blood is pumped around the body
    b   animals create offspring (babies)
    c   the body changes during adolescence
2   Animals that lay eggs are:
    a   puberty
    b   viviparous
    c   oviparous
3   True or false?
    a   In puberty, a boy's testicles get larger.
    b   Girls grow an Adam's apple.
4   Match the words to the correct sentences.
    a   Ovary   When the egg has been fertilised by the sperm.
    b   Sperm   Where the eggs form in a female's body.
    c   Pregnancy   Made in the male's testicles.
5   Give three ways your body will change during puberty.
6   Name three of the same ways that girls' and boys' bodies will change during puberty.
7   Use the following words to copy and complete the sentences.

    (12 and 16)   (hormones)   (same)   (puberty)

    (adolescence)   (10 and 14)

    _____ is the time during _____ when a boy or girl's body

    begins to change and develop as they become an adult. Puberty is the time when humans reach

    sexual maturity. This process usually takes place between the ages _____ for

    girls and the ages _____ for boys.

When the body is ready to begin puberty; it releases special _____

(chemicals that tell the body what to do), these hormones change different parts of

the body depending on whether you are a boy or a girl, some of the changes are the

_____ for both.

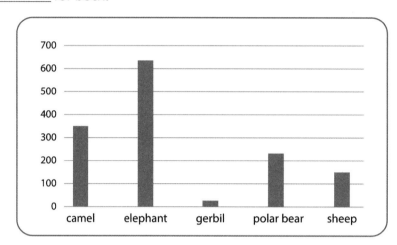

8   The graph shows the gestation period of different animals; however Stage 6 learners did not complete the graph.

a   Write a title for the graph.

b   What should the label be on the *x*-axis?

c   What should the label be on the *y*-axis?

d   Which animals have a longer gestation period than sheep?

e   Which animal has the shortest gestation period?

f   If the gestation period for a rhino is 450 days, where should the bar be on the graph?

Food chains

Do you remember?

Food chains

A **food chain** is a diagram that shows how organisms in a habitat feed on one another. Arrows are used in a food chain to show the direction that energy flows through the food chain. In a food chain, there are producers and consumers. **Producers** are organisms that make their own food, such as plants. **Consumers** in a food chain are living things that eat other organisms (living things).

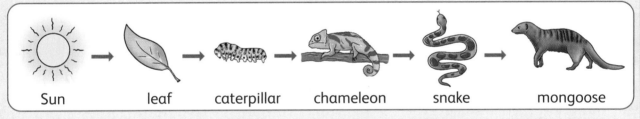

| Sun | leaf | caterpillar | chameleon | snake | mongoose |

Look at the diagram of the food chain. Read the food chain, each time you see the arrow say 'Energy is transferred from one organism to its consumer'. Then write the story that the food chain is telling. Begin with:

Energy is transferred from the Sun to the leaf.

_____

_____

_____

Write down which organisms are consumers and which are producers.

_____

_____

Activity 1

Look at the food chains below, write the story of each food chain.
For each food chain list the consumers and producers.

A

Producers:_____

_____

Consumers: _____

_____

Story of food chain: _____

_____

B

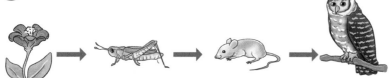

Producers:_____

_____

Consumers: _____

_____

Story of food chain: _____

_____

Activity 2

Draw the food chain for the organisms below. Use arrows to show the flow of energy through the food chain. There is an organism missing. Draw an organism to complete this food chain.

Activity 3

You are going to make three mini concertina books (horizontal or vertical) to show a food chain for each of the following habitats:

- Ocean
- Desert
- Where you live, for example, your own garden, or somewhere in your local area.

You will need:
- paper or card
- scissors

You will need to research each habitat and work out a food chain for each one.

When you have completed each book, put it on display for others to read.

Food webs

Do you remember?

Food webs

A **food web** is made up of food chains linked together. A food web shows how the living things in a habitat rely on one another for food. It looks more complex than a food chain because it shows many food chains, the connections between food chains and the different paths of energy.

Food webs show the feeding relationships (links) between different living things and the energy transfer from one living thing to another.

Activity 1

Look at the ocean and desert food webs below.

How many food chains can you find? On a separate sheet of paper, draw each food chain that you can find. One for the ocean and one for the desert have already been done for you.

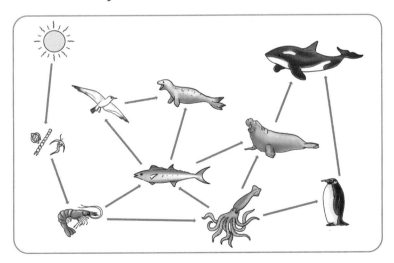

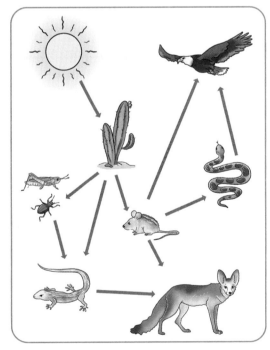

Sun → phytoplankton → krill → fish → leopard seal → whale

Sun → desert plant → insect → lizard

Activity 2

Look at the two food webs on page 33, are there any organisms that you do not know?
Research those organisms and write a fact card with three facts for each one.

Activity 3

Revision approach

Key word cards

In Unit 2, you made key word cards to help you revise and remember key scientific words.

You are going to make a set of key word cards for the scientific words below to help you revise
and remember words connected to food chains and food webs.

Draft your key word cards first and ask someone to check them before you make the cards for
your collection.

(food chain)   (food web)   (toxic)   (producer)   (consumer)   (energy flow)

Revision approach

Model answer

A model answer is an ideal response to a question.

Looking at different answers to questions and thinking about how they can be improved using
your own knowledge is one way of helping you to revise and remember scientific knowledge.
Thinking about the strengths and weaknesses of an answer is a way of assessing how good an
answer is, we can then rewrite it to improve it.

Activity 4

Stage 6 learners were asked to explain the difference between a food chain and a food web.

Here is what one of the learners wrote:

*The difference between a food chain and a food web is that a food chain shows animals that eat a
plant and which animals are dependent on the other for food. A food web shows lots of food chains.*

1 Look at the key word cards that you made in Activity 3 and compare the words used above with
   your definitions.

2 How good do you think their answer is? Explain your answer.

   Looking at this explanation I really like …

   _____

   _____

   This is what I would have written to improve the learner's answer …

   _____

   _____

Science in context

The Minamata Story

Toxic materials in ecosystems

Toxic materials are substances that can be poisonous to living things, plants, and animals including humans. This is the story of toxic materials that polluted the sea in Minamata Bay, Japan which led to people becoming ill and many dying.

What happened?

A local factory was releasing waste into the sea at Minamata Bay. The waste contained a substance called mercury. Mercury is metal that is liquid at room temperature and is found in rock, soil and water throughout the world. Human activities, such as some manufacturing industries that use mercury, can increase the amount of mercury in the environment through pollution.

When mercury enters the sea, bacteria change into a very toxic type of mercury called methylmercury. Although the amounts of mercury being released into the sea can be quite low, its impact on the marine food chain can be high.

How did it affect the food chain?

The methylmercury (say meth - il - mer - cure - ee) entered into the food chain through the seawater which was absorbed by microscopic algae called phytoplankton, these were then eaten by tiny marine animals called zooplankton. The methylmercury continued to move through the food chain because the phytoplankton were then eaten by shellfish and lots of different small fish.

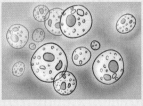

phytoplankton

zooplankton

The small fish were then eaten by larger fish such as tuna, shark, swordfish and marlin.

As a result, these larger fish ingested (ate) large amounts of methylmercury, and because larger fish live longer (for example, a shark can live up to 30 years, a tuna fish up to 40 years) the methylmercury accumulated (collected) in these fish over many years. Therefore, there was more methylmercury in the top predator fish, sharks, marlin and tuna.

When toxins accumulate in an animal over time scientists call this **bioaccumulation**. The methylmercury collects in the muscles of fish. Scientists have found that the top predators can have 10 million times the amount of methylmercury in their muscles than amounts found in their habitat.

How did this affect humans?

Humans, of course, are part of this food chain because they catch and eat tuna, sharks and marlin, which means that many humans ate fish that was toxic (poisonous). In May 1956, four people from Minamata died, the scientists did not know why. Now scientists understand what happened. The factory continued to pollute the sea until 1968 and the Japanese Government states that 2 955 people suffered from fever, convulsions (spasms), numbness in the hands and feet, muscle weakness, damage to hearing and speech and many went into a coma. Overall, 1 784 people died from eating contaminated fish and shellfish.

Is methylmercury still a problem today?

Today methylmercury still enters the marine food chain because mercury is a natural substance. The amount has increased because of humans burning fossil fuels like coal and activities such as gold mining. Scientists advise people not to eat more than one portion of shark, marlin and tuna per week. Pregnant women and children are advised that they should not eat shellfish, shark, swordfish or marlin.

Revision approach

Rich picture poster

A rich picture is a way of showing a complex situation using pictures, diagrams and individual words, phrases and colour coding.

Look at the rich picture poster that someone has started to help them remember the story of Minamata Bay and how toxic substances moved through the food chain and affected sea life and humans.

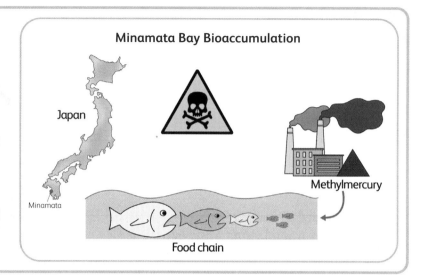

Activity 1

Use a large piece of paper and create your own rich picture using the information from the newspaper article on page 35.

Think about the following:

- What are the key ideas in the story?
- What facts should be included?
- Do you need to use arrows?
- How can a food chain diagram show part of the story?
- How can your rich picture help someone else to revise the science behind the Minamata story?

Share your rich picture with a partner, ask them to say what they have learnt from your poster. How well do they think your rich picture helped them?

When you have finished, you should be able to use it to answer these questions.

1   Identify the toxic substance that was released into the sea by human activity.

2   Describe how the toxic substance moved through the food chain to humans.

3   Explain why the top predators in the habitat were most at risk from the pollutant.

4   How can we use our learning from this to stop similar things happening in the future?

Revision quiz

Revisit the work that you have done in this unit. Read your key word cards to help you remember scientific words and information. When you are ready complete this short quiz.

As you work through the quiz, you can help yourself by:

- Reading each question carefully – make sure that you understand the question.
- Look for key words, use them in your answer.
- Answer the question in your mind first before you write it down.
- If the question asks for an explanation, jot down key words that you will need to use, check that you use them correctly in your explanation.

Answer the questions on a separate sheet of paper or in your notebook.

1 In a food chain a producer is …
   a where people make something
   b an organism that makes its own food like plants
   c an organism that eats other animals

2 'Toxic' means
   a an animal   b edible   c poisonous

3 This food chain is incorrect. State what is wrong with it and then write the correct food chain.

     Sun   eagle   grass   rabbit

4 In a food chain, a consumer is …
   a a tree
   b an organism that does not make its own food, instead it eats other organisms
   c someone who buys things to eat

5 Name two similarities between a food chain and a food web.

6 Name two differences between a food chain and a food web.

7 Explain how a toxic substance like mercury is passed along the food chain.

8 Use the food web below to answer the following questions.
   a On a separate sheet of paper, draw three food chains from the food web.
   b Name three predators.
   c Name three consumers.

9 On a separate sheet of paper, draw a food web to show how the toxic substance, methylmercury, can poison humans.

10 Use the word 'bioaccumulation' to explain why a tuna fish might have 10 million times the amount of methylmercury in their muscles than in their habitat.

11 Explain why top predators are most at risk from pollutants in an ecosystem.

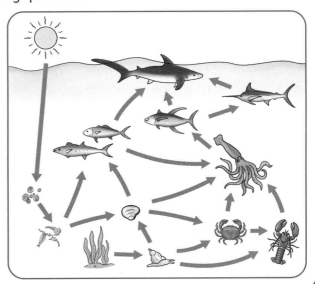

This unit will help you to revise your learning of:

Properties of materials
- everything has matter, including gases
- a property of a substance is the temperature at which it changes state
- some materials are thermal conductors, this is a property of the material.

Changes to materials
- physical changes are reversible (can be changed back to what they were before)
- chemical changes are irreversible (cannot be changed back to what they were before)
- boiling and evaporation are different processes
- temperature can affect how a solid dissolves; and we can use the particle model to describe how this happens
- in a chemical reaction substances called reactants act together to form new substances, these are called products.

 Pages 42–67

Revision approach

What are revision hexagons?

Revision hexagons are hexagon shapes that tesselate (link together). The ideas and facts written in each hexagon must link to what is written in the hexagons around it.

This revision approach helps you to:
- think about what you know
- recall facts and ideas
- make links between learning.

Look at the example. The word 'Materials' has been placed in the centre. All the other hexagons link to the 'Materials' topic. The rule is that the information that is written in each hexagon must be about materials, and each hexagon must link to the hexagons around it.

As you complete each hexagon, this approach will help you to think of other ideas and make another link. You will be surprised how much you know and how many of the different ideas about materials are linked together.

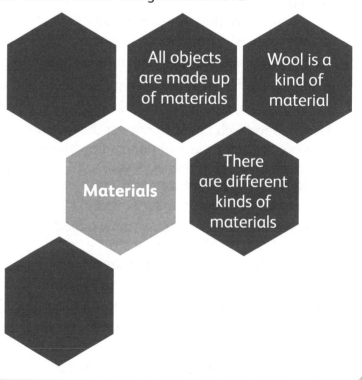

Activity 1

Hexagon page

Work with a partner. Help each other to remember what you both know about materials and make a hexagon page each. Your teacher will give you a hexagon page template to work on. It does not matter if you write the same thing, as long as you know and understand what you are putting in each hexagon. You can add to and change hexagons that you have already written, so it is useful to write them in pencil and not pen.

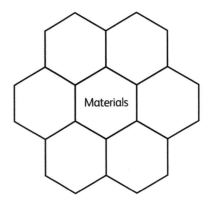

Using your page of linked hexagons, write the word 'Materials' in the centre hexagon.

Think about everything you know about materials. Fill each hexagon with a fact or idea that you remember about materials, you can use words, phrases and even pictures and diagrams. Remember that each hexagon must describe something about materials and the information in each hexagon must link to the ones around it.

Challenge yourself to fill as many hexagons as you can.

Hint

Look around you at different everyday objects and their materials. Revisit pages 42–67 in Unit 4 of the Learner's Book to refresh your memory.

Activity 2

When you have finished, read through your hexagon page. Is there anything that you want to change or are there more ideas or facts that you want to add? If you are working with a partner, compare their page with yours. Check each other's work to make sure that what you have written is correct.

Materials

Do you remember?

Materials

Everything is made up of matter, matter is anything that has volume (takes up space) and has mass. Having mass is a property of a material.

Materials are different types of matter from which objects are made, for example, wood, metal, glass, fabric.

Materials can be in one of three states:
solid, for example: rock, wood   **liquid**, for example: water, lemonade
gas, for example: carbon dioxide, oxygen

The different states of matter have different properties.

Solid	Liquid	Gas
Keep their shape	Can be poured	Can be poured
Can be cut or shaped	Take the shape of their container	Usually invisible
		Move around

A material can change from one state to another.
solid → liquid → gas   gas → liquid → solid
Heated ──────────→   Cooled ──────────→

Activity 1

1   Stage 6 learners weighed two 500 ml bottles of water. One bottle contains still water, and the other bottle contains sparkling water. Sparkling water has bubbles of carbon dioxide in it.

   a   Which bottle of water do you think was the heaviest?

   _____

   b   Explain why you think that.

   _____

   _____

   c   Explain how this activity shows that gas has mass.

   _____

   _____

   _____

2   Devise a test to find out how much carbon dioxide gas is in the sparkling water. Share your test with a partner, do they think your test will work? Why?

   _____

   _____

   _____

   _____

   _____

   _____

   _____

Activity 2

Draw lines to match the word to the correct definition.

Word
A   Condensation
B   Evaporation
C   Freezing
D   Melting

Definition
1   The process that happens at the surface of a liquid when it changes state from a liquid to a gas.
2   The process in which a material changes from a liquid to a solid.
3   The process in which a material changes state from solid to liquid.
4   The process in which a material changes state from gas to liquid.

Activity 3

Read the information in the **Do you remember?** panel on page 39 and look back at the activities you have carried out.

1   List the key scientific words linked to materials, check to see if you used them in your hexagons, if not, you can now add them.

2   Check the ideas in the **Do you remember?** panel, did you remember all of these ideas? If not, add them to your hexagons.

Changing states

Do you remember?

Changing materials

Changing state is a property of substance. Substances change state, usually when they are heated or cooled, for example, liquid water is a substance (material), it turns into water vapour (a gas) when it is heated enough. When liquid water is cooled enough, it changes into ice which is a solid.

Water particles have different arrangements and move differently when water changes state.

Even though the particles of water are arranged differently when it is a liquid, solid or gas, the particles themselves have not changed, they are still the same, what has changed is how they are organised and move.

Remember, water also has mass, even when water changes from one state to another its mass does not change. When 1 g of water freezes it forms 1 g of ice; when 1 g of water boils it forms 1 g of water vapour.

Remember that the temperature at which a substance changes state is a property of the substance.

Here are important numbers that you should remember about water as a substance.

Water as a liquid changes state to ice – a solid at 0°C (32°F), this is known as its freezing point.

Water as a liquid turns to water vapour – a gas at 100°C (212°F), this is known as its boiling point.

Activity 1

1   Read through the information above and list three things that you already knew.

_____

_____

_____

2   List something that you did not know or were unsure about.

_____

_____

3   List the key words that you need to remember.

_____

4   Go back to your hexagons, check to see if you have included any of the ideas from the **Do you remember?** panel. Add any information that you think should be included in your hexagons to help you remember ideas about materials and their properties.

Activity 2

1 Identify which diagram shows a solid, liquid and gas.

 (A)    (B)   (C)

_____   _____   _____

2 Explain how you know what state each diagram represents.

_____

_____

3 Match these statements to the correct diagram.

Statement	Diagram (A, B, C)
Particles randomly arranged but close.	
Particles have regular arrangement and are close.	
Particles randomly arranged and far apart.	
Particles move slightly (vibrate) but stay in the same position.	

Do you remember?

Evaporation

Evaporation is a reversible change that takes place on the surface of a liquid. The liquid changes state to a gas.

Think of a puddle of water, over time the puddle becomes smaller because the water in the puddle changes state to become a gas – it evaporates.

Imagine the particles of water in the puddle, as they are heated by the Sun the particles are given more energy. The particles at the surface get more energy and move the fastest. These particles move fast enough to be able to change into a gas and move into the air. When this happens, it is called **evaporation** and the water becomes a gas called **water vapour.**

Remember, evaporation only takes place at the surface of a liquid. So, evaporation only takes place at the surface of the puddle, and this happens as the puddle is being heated.

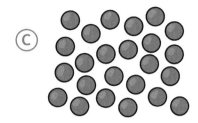

When a liquid boils, all the particles in the liquid are changing into a gas. If a liquid boils all the particles have enough energy to form bubbles of gas in the liquid. Boiling takes place throughout the liquid, and not just on the surface. This only happens at the boiling point of the liquid, and the liquid does not heat up anymore. Different liquids have different boiling points. Water boils when it reaches 100°C.

evaporation   condensation

Condensation is the reverse of both evaporation and boiling. It is when the gas is cooled enough that it changes state from a gas to a liquid, this is a reversible change.

Activity 3

On a separate sheet of paper, draw a Venn diagram to show the similarities and differences between evaporation and boiling.

Activity 4

Fact file cards

You are going to revise reversible and irreversible changes by creating a set of fact file cards, like the one in the picture. This revision approach will help you check that you understand examples of reversible and irreversible changes.

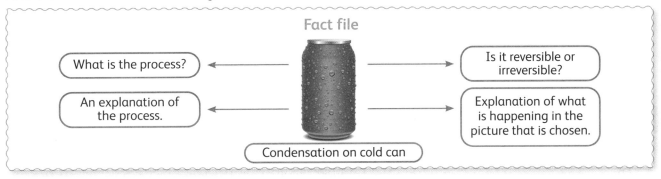

Fact file

What is the process?

Is it reversible or irreversible?

An explanation of the process.

Explanation of what is happening in the picture that is chosen.

Condensation on cold can

Make a fact file card for each of these processes:

(Melting) (Evaporation) (Boiling) (Freezing)

When you have finished, swap your fact file cards with a partner. Do you think their fact file card is correct, explain why? What could you suggest, to improve, or add to their card?

Revision approach

Thinking and Working Scientifically

Working from a graph

In this set of activities, you are going to use data from an investigation. The graph shows the results of a fair test investigation. You are going to use the results in the graph to work out how the investigation was planned and carried out, and think about the reasons for what the learners did. This is a useful way of revising what you can remember about Thinking and Working Scientifically.

Activity 5

Here are the results of an investigation about evaporation. Use these results to revise Thinking and Working Scientifically.

1 Before you begin, look at the graph. Tell yourself or a partner the story of the graph. What does the graph tell you? What do you think the learners who created the graph did to get the data for the graph? Is your story the same as your partners? Does the story fit with the data shown on the graph?

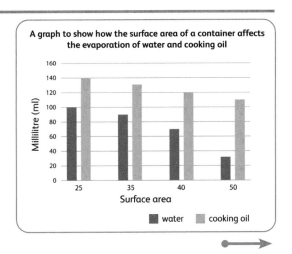

A graph to show how the surface area of a container affects the evaporation of water and cooking oil

43

2   Look at the graph. Write the question that the Stage 6 learners were investigating.

3   Use the data to think about how they kept their test fair to collect this data.

4   Copy and complete the memory map plan for the investigation the learners carried out.

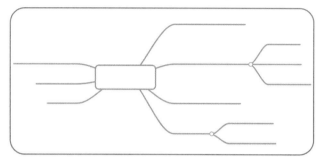

5   Make sure that you identify the independent, dependent and control variables.

Activity 6

Use the data in the graph and your plan to carry out the following activities.

1   On a separate sheet of paper, create the table of results that the learners used to collect the data in the graph.

2   Use the data to answer the question that the learners were investigating.

3   What do you think the learners could have done to make the data more reliable?

4   Write a sentence to explain how the graph can be used to show that the water and cooking oil evaporated.

5   Name five other liquids that you could use to compare the rate of evaporation with oil and water. Make a list, explain why you chose those liquids.

6   Complete the table. Using the data in the graph, make predictions about whether the evaporation rate of the liquids that you have chosen would be slower or faster than oil or water and give your reasons.

Liquid	Prediction	Reason
1.		
2.		
3.		
4.		
5.		

Dissolving

Do you remember?

Dissolving

When some solids with small particles mix with a liquid, for example, water, they dissolve.

The liquid used to dissolve a solid is called a **solvent**.

Look at the diagram. When a substance dissolves, it might look like it has disappeared, but it has mixed with the water. This liquid is called a **solution**.

Soluble substances are substances that dissolve in water.

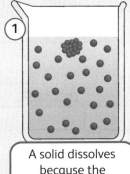 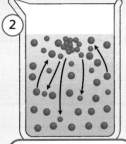 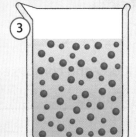

A solid dissolves because the particles of the solvent collide with the particles of the solid.	The solid particles gradually move away from each other until they are evenly spread through the solvent.	The solid particles are still in the solution. They are just spread out and you cannot see any particles of the solid.

Substances that do not dissolve in water are called **insoluble substances**.

When a solid is mixed with a liquid but does not dissolve in it, the liquid becomes cloudy, this is called a **suspension**. Flour forms a suspension, flour is insoluble.

Activity 1

A Stage 6 learner watches the teacher mix a teaspoon of sugar into a glass of water. The learner says, 'The sugar has disappeared in the glass.'

1   Why is this statement not correct?

_____

_____

2 What has the learner not fully understood about dissolving?

_____

_____

3 How could you demonstrate or explain to the learner what has really happened?

_____

_____

_____

_____

Activity 2

1 Use the information from page 45 to sort these substances into soluble and insoluble.

coffee   sugar   sand   salt   pasta   pebbles   vitamin tablet   rice

Copy and complete this table.

Substance	Soluble	Insoluble	Unsure

2 Look at your table. Were there any substances that you were unsure about? If so, you could try to dissolve them in some water to find out or research them.

Activity 3

Stage 6 learners carried out an investigation to answer this question:

How does the temperature of a solvent affect how quickly a solid dissolves?

They started their plan but did not complete it. Copy and complete their plan to answer the question.

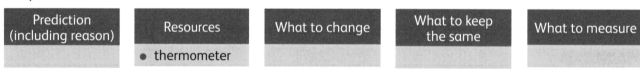

Prediction (including reason)	Resources	What to change	What to keep the same	What to measure
	• thermometer			

1 On a sheet of graph paper, sketch what the line graph should look like for this investigation, remember to:
   ○ label the axes
   ○ use different colours for each of the different temperatures.

2 For each temperature you have chosen in the investigation, draw a particle diagram to help explain your prediction. Show how the temperature of the solvent affects how quickly the solid dissolves.

Activity 4

Look back over your work so far.

What else can you add about changing materials to your hexagons? Make sure that the information you put in each new hexagon links to the hexagons around it.

Reversible and irreversible changes

Do you remember?

Reversible and irreversible changes

An irreversible change cannot be easily reversed (changed back), for example, when you bake a cake you cannot get the flour, butter and sugar back.

When an irreversible change happens there are three scientific terms that you must remember.

Chemical change: An irreversible change.

Reactants: The materials that react (act together) and change.

Products: In a chemical change one or more new materials are usually formed. These new materials are called products.

Think about how you can remember these words, for example, for the word 'product', you could think about the word 'produce' – to make something, the product is made (comes after). This will help you to remember that the reactants come first.

Activity 1

Stage 6 learners were taught to use these signs to show reversible and irreversible changes.

→ Irreversible

←→ Reversible

1   What do the arrows mean? Why have they been used? Explain.

_____

_____

2   Make up your own hand sign to help you remember the words 'irreversible' and 'reversible' and the written arrow signs.

Activity 2

Copy and complete the table below. Use the signs to show which processes are reversible and irreversible. The first one has been done for you.

Process	→ or ←→	Describe two changes that take place.	Why is the change → or ←→?
Burning match	→		The match has changed, it is black, smaller and cannot be changed back to before it was lit.
Making an ice lolly			
Melting chocolate			

Process	→ or ↔	Describe two changes that take place.	Why is the change → or ↔?
Rusty iron nail			
Condensation on a mirror			
Dissolving salt			
Baking a cake			
Toasting bread			

Activity 3

Here are four irreversible changes. Copy and complete the table.

The product is the new material that is formed, the smoke and ash.

A burning wooden log is a chemical change – it is irreversible, you cannot get the log back.

The wood is the material taking part in the chemical reaction – the wood is the reactant.

Irreversible change	Reactant	Product
Bread baking		
Egg frying		
Toasting bread		

Activity 4

Try this activity.

You will need:
- bicarbonate of soda
- jar
- 1 plastic glove
- measuring jug
- measuring spoons
- vinegar

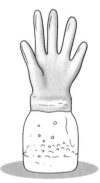

Stage 6 learners did the following:
- Poured 30 ml vinegar into the jar.
- Poured 15 ml bicarbonate of soda into the fingers of the glove.
- Carefully put the glove onto the jar as shown in the picture.
- Held the glove by the fingertips and shook them so that the bicarbonate of soda fell into the jar.
- Observed the results.

Carry out the same activity as the learners. Use your observations to answer the following questions. Write the answers in your notebook.

1   Was this a reversible or irreversible reaction? Explain why.

2   Is this a chemical change? Explain why.

_____

_____

3   Name the reactant in this activity.

_____

4   Name the product in this activity.

_____

5   Carbon dioxide was produced, what happened in this activity that would help to prove that carbon dioxide is a gas?

_____

_____

6   On a separate sheet of paper, draw a diagram of what happened inside the jar and glove when there was a chemical reaction between the bicarbonate of soda (solid) and the vinegar (liquid).

Activity 5

How will you remember the three scientific words: **chemical changes**, **reactant**, and **product**, and the meaning of each one?

Look at each of the ideas provided and decide which approaches you will try to help you remember these words.

Ask a partner to test you a day later, two days later and a week later. Can you still remember all of the words?

If you forget any of the words, try a different approach to help you.

Read, read, read, keep repeating the word to yourself or a friend.

Use the new words in sentences.

Match the word to a picture in your head.

Make key word cards.

Split the words up to remember them, for example, pro   duct.

Create a hand signal for each word.

Create a memory place, for example, fire – chemical reaction – sitting by a log fire at home.

Link the word to a drawing.

Activity 6

Look back over your work so far.

What else can you add about changing materials to your hexagons? Make sure that the information you put in each new hexagon links to the hexagons around it.

Thermal conductivity

Do you remember?

Thermal conductivity

Another property of different types of materials is how well the material lets heat pass through it, this is called thermal conductivity. In this section, you are going to revise your scientific knowledge of **thermal conductors** and **insulators**.

Thermal insulators are materials which do not conduct heat very well and so we can use them to stop heating or cooling happening, this can help us keep things hot or cool, for example, polystyrene, plastic, wood, air and fabric.

Thermal conductors are the opposite of thermal insulators, they are materials that allow heat to move through them easily. Metals are very good thermal conductors. We can use thermal conductors to transfer heat, for example, metal saucepans transfer heat so that we can boil water.

Activity 1

1   Explain the difference between thermal insulators and thermal conductors.

_____

_____

2   Make a list of four materials that are thermal insulators and one thermal conductor.

_____

_____

Activity 2

Draw a Venn diagram or use the template from your teacher to sort the following kitchen objects according to whether the materials they are made from are thermal insulators, thermal conductors or both.

(wok with wooden handle)   (plastic spatula)   (fabric oven gloves)   (plastic bottle)   (metal frying pan)

Activity 3

1   Look at the picture. What do you think the Stage 6 learners were trying to show in this activity about thermal conductivity?

_____

_____

2   Predict what the results should be and explain why for each result. Copy and complete the table below.

Spoon material	Prediction	Reason

⚠ Do not use boiling water

wooden spoon → | plastic spoon → | metal spoon →

hot water   hot water   hot water

Science in context

Global warming

People across the world are talking about and acting on climate change. In colder countries, an important topic of discussion is how to reduce the amount of energy people use to heat their homes.

Heat travels from higher temperatures to lower temperatures, so when we heat our homes, heat will transfer from any uninsulated area to the cooler temperature outside.

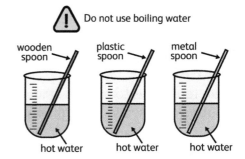

A brick house with poor insulation will lose over a third of its heat through the walls, one-quarter through the roof and the rest through the doors, windows and floors.

In colder countries insulating homes is very important. The less a home is insulated the more heat transfers to the cooler temperature outside, so people use more electricity, gas, wood or oil to heat the home to keep them warm.

A well insulated house keeps the heat in and so people stay warm, they use less electricity, gas, wood or oil to heat the home, reducing the energy they use.

We know different materials have different properties, some materials for home insulation are better insulators than others.

Many people are trying to use natural, sustainable materials to insulate their homes, here are some materials being used, they might surprise you.

Denim jeans

Unfortunately, tons of clothing end up in landfills and it takes years for it to decay. One way to help is to send denim jeans for recycling, where they will be shredded into small pieces and used to insulate homes. It is a good thermal insulator so it will help to ensure that heat is not transferred from a home into the surrounding environment.

Sheep's wool

Sheep's wool keeps the animal warm in cold climates, so it is not surprising that sheep's wool is a good thermal insulator and is used

as insulation for homes. Wool is a sustainable material, once the sheep have been shorn, the wool grows again.

Straw

Would you believe that straw is a good insulator? Straw has good thermal insulation properties. Straw is the dry stalks of cereal plants such as wheat, oats and barley. People use straw bales to insulate the walls of houses ensuring that they stay warm in colder winter months and cooler in hotter summer months.

Activity 1

Create four slides to explain what thermal insulators and thermal conductors are and how thermal insulators are used in the fight against climate change.

Activity 2

Prefixes

Remember that prefixes are a group of letters that change the meaning of a word when they are added to the start of a word.

All of these words use the prefix 'therm'.

(thermographic scanners) (thermal imaging) (thermogram) (thermographer)

Answer the following questions.

1 What does the prefix 'therm' mean?

_____

2 What do each of the words above mean?

_____

_____

_____

_____

Here is a picture taken by a thermal camera. The different colours on the thermogram show different temperatures: Coldest areas: Blue; Warmest areas: Red, orange and yellow

3 Look at the picture. Where does this house need insulating? Explain why.

_____

_____

4 Where do you think the house already has insulation? Explain why?

_____

_____

Revision quiz

Revisit the work that you have done in this unit, check your revision hexagons, read through them to remember scientific words and information. When you are ready complete this short quiz.

As you work through it you can help yourself by:

- Reading each question carefully – check that you understand the question.
- Look for key words, use them in your answer.
- Answer the question in your mind first before you write it.
- If the question is multiple choice, decide which answers you know are definitely incorrect, then think about the answers that are left, to decide which one is correct.
- Check your answers to make sure that you do not want to make any changes.

Answer the questions on a separate sheet of paper or in your notebook.

1   Which word describes substances that dissolve in water?

   a   insoluble   b   soluble   c   solute

2   Which of these substances will dissolve in water?

   a   rice   b   salt   c   pasta

3   The arrows showing reversible and irreversible change is incorrect. Draw the correct arrows.

   bread dough ← → bread

4   A thermal insulator

   a   is a material that does not let electricity pass through.

   b   is a material that is waterproof.

   c   is a material that lets heat pass through easily.

   d   is a material that does not let heat pass through easily.

5   Name two similarities between boiling and evaporation.

6   Explain how the temperature of the water affects how quickly a substance dissolves.

7   Name three different solutions.

8   Which statement is correct?

   a   Boiling is a reversible change that takes place on the surface of a liquid where the liquid changes state to a gas.

   b   Condensation is a reversible change that takes place on the surface of a liquid where the liquid changes state to a gas.

   c   Evaporation is a reversible change that takes place on the surface of a liquid where the liquid changes state to a gas.

9   Name three irreversible changes.

10 Match the correct definition to each word.

   a   Reactants   In a chemical change one or more new materials are usually formed.

   b   Chemical change   The materials that take part in and change during a chemical change.

   c   Products   An irreversible change.

11 What could you do to make a substance dissolve faster in water?

   a   cool it down   b   freeze it   c   heat it up

What will you learn

This unit will help you to revise your learning of:

- mass is measured in kilograms (kg), and weight is measured in newtons (N)
- gravity is a force. When gravity changes, the mass of an object stays the same but the weight changes.
- the mass and shape of an object can affect if it floats or sinks
- objects weigh less in water than they do in air
- we use force diagrams to show the type,
- size and direction of forces acting on an object
- different forces have different effects on an object at rest and in motion (moving)
- when balanced forces are acting on an object, it will not change its motion, shape or direction
- a fluid exerts an upward force, called upthrust, on any object in it.

 Pages 68–80

Forces

Revision approach

Create forces key word cards

You have used key word cards in previous units. Key word cards help you to remember the most important words and ideas linked to that word.

For this unit, make 'forces' key word cards that use images. Images are really useful and can help you remember things better.

Look at the key word cards made by Stage 6 learners below.

There must be two objects that touch. These can change the motion, direction and shape of the object they act on. They can be balanced or unbalanced.	**Contact force** 
There must be two objects that interact, these objects are not touching. Examples of non-contact forces are weight (gravity) and magnetism. They can be balanced or unbalanced.	**Non-contact force** 

Activity 1

This is a list of the words that you need to learn and remember about forces.

You will need:
- small pieces of card or plain paper
- coloured pens

(contact force) (non-contact) (motion)

(force diagram) (interacting objects) (support force) (balanced forces)

(steady speed) (unbalanced forces) (mass) (weight) (gravity)

(sinking) (gravitational attraction) (newtons) (floating) (upthrust)

1   Write each word on the front of a card and add any images that will help you remember what the word means.

2   On the back of your card write a definition. Check your definition using a dictionary. Think about whether your definition is a good one or if parts of it need changing. Add as many other ideas as you can that link to the key word.

3   Are there any words you do not know or find hard to write a definition for? Spend time finding out more about this word; use the Learner's Book, if you have one, to help you with this. Ask a partner to check your key word card to check your thinking.

4   Test yourself by only looking at the key word on the front of your card. Can you remember what is written on the back? Ask a partner to test you to see if you can remember everything that is written on the back.

Do you remember?

Force arrows

Force diagrams show the forces that act on an object. Force diagrams use arrows to represent forces. Force arrows must meet these four rules:

1   They must show the direction in which the force is acting. This is shown by the direction the arrow is pointing.

2   They must show how strong each force is. The longer the arrow, the stronger the force.

3   They must show where the force is acting from. The base of the arrow (the flat end) shows us where the force is acting from.

4   They must be labelled with the name of the force, including the two objects that make up this force.

Activity 2

There are four different rules that you need to remember when drawing force arrows.

1   On separate cards or sticky notes write out each of the rules.

2   Place the four cards in front of you for 20 seconds and try to remember each rule, then collect the cards.

3   Now deal out only three of them, which is the one that is missing? If you forget look at your last card. Collect them again and shuffle.

4   Now deal out only two cards, which are the two that are missing? Collect them again and shuffle.

5   Can you say the four rules without looking at any cards?

You will need:
* sticky notes, pieces of card or paper
* pen or pencil

Activity 3

Force diagrams use force arrows to help us represent the different forces that are acting on an object. It is important that the arrows are drawn carefully and that they are labelled fully.

The picture on the right is a good example of a force diagram. Imagine that you are a teacher and you have to mark this diagram. With a partner, discuss what reasons you would give to mark it four marks out of four?

support force of the table on the cat

weight of the pull of the Earth on the cat

Activity 4

Sometimes people make mistakes! Some Stage 6 learners tried to improve the force diagrams below:

(A)

Sailboat: push force of
wind, push force of water

(B)

Seed: up force,
down force

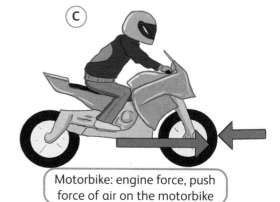

(C)

Motorbike: engine force, push
force of air on the motorbike

1 Can you spot the errors in each diagram?

2 Once you have spotted the errors in each diagram, write a feedback comment to the learner to
help them not make the same mistake in the future.

A _____

B _____

C _____

Balanced and unbalanced forces and motion

Do you remember?

Balanced and unbalanced forces and motion

Motion is the way an object moves, and forces can affect this. If forces are balanced, they
do NOT change the motion of the object. This does not mean that if forces are balanced the
object cannot be moving.

If the forces acting on an object are balanced, it will either stay still or it will keep moving at the
same rate. We call the same rate, **steady speed**.

Activity 1

A Stage 6 learner does not understand how there can be forces acting when an object is
not moving.

All of the named objects in the pictures below involve balanced forces. Draw and label the force
arrows on the pictures below, to show the learner that there are forces acting even though the
object is not moving.

(1) child

(2) teddy

(3) child

Do you remember?

Unbalanced forces and motion

If forces are unbalanced, they DO change the motion of the object. If forces are unbalanced, it means that the force acting in one direction is greater than the force acting in the opposite direction. Unbalanced forces can either slow down and stop or speed up and start the motion of the object.

Activity 2

Many people find it difficult to understand that an object can have balanced forces acting on it **and** be moving.

If the object is moving and the forces are balanced then we know the object CANNOT get any faster or any slower, unless the forces acting on it change. If the object is moving and the forces are unbalanced, we know that the motion of the object WILL change.

1   Look at the following examples.

　　A   A person who is running around a racetrack at a constant 5 miles per hour.

　　B   A motorbike rider who is pulling their brakes so they can stop.

　　C   A swimmer who keeps going at a constant speed of 2 miles per hour.

　　D   A skateboarder on a flat road going at a constant speed of 10 miles per hour.

　　E   A dog that jumps up to run after a ball.

　　F   A cyclist who is freewheeling down a hill.

　　a   Which of the examples of moving objects are those which have balanced forces? How do you know?

　　　　_____

　　　　_____

　　b   Which of the examples of moving objects are those which have unbalanced forces? How do you know?

　　　　_____

　　　　_____

2   Make up three examples of your own, two that involve balanced forces and one that does not. Test another person and see if they can spot the unbalanced forces example.

　　i   _____

　　ii   _____

　　iii _____

3   Make up a rhyme to help you remember when there are balanced forces acting on a moving object. Write your rhyme and share it with a partner.

　　_____

　　_____

4   Work together to produce actions that go with your rhyme. Your actions should help people who are watching understand what you are teaching them about balanced forces.

Do you remember?

Double bubble

A double bubble is another type of memory map. A double bubble is a useful way to check that you have understood ideas. The double bubble helps you look at two different ideas and compare the similarities and differences.

The centre of the large circles contains the two ideas you are thinking about. In the middle of the double bubble, in the blue hexagons are the similarities. Here each hexagon will contain one thing that is true (the same) for both ideas. In the outside circles are the differences. Here each circle will contain one thing that is different for each of the ideas you are thinking about.

Using this type of memory map will help you to make connections and remember what you have been learning about.

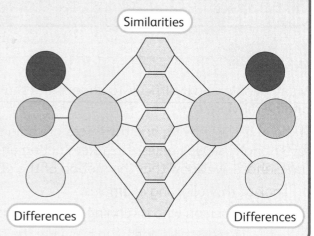

Activity 3

Creating a double bubble for balanced and unbalanced forces

You will need:
- plain paper
- coloured pens
- double bubble template

1   On a separate sheet of paper, draw out a double bubble or use the double bubble template from your teacher.

2   In the large left circle write 'Balanced forces' and in the large right circle write 'Unbalanced forces'.

3   Write as many ideas as you can in the middle hexagons that are the same for both 'Balanced forces' and 'Unbalanced forces', for example, they involve two objects interacting. Add extra hexagons if you need to.

4   In the outside circles make a note of ideas that are different in 'Balanced forces' and 'Unbalanced forces'. For example, in the purple circles you could write for 'Balanced forces', *objects do NOT change their motion*; and for 'Unbalanced forces', *objects DO change their motion*. Add extra circles if you need to. Try and make the coloured circle ideas connect. In the example above the person is thinking about the motion of the objects.

5   Ask a partner to check your double bubble to see if you have got the ideas correct.

6   To test your memory:

   a   Cover the middle hexagons and see if you can remember all of the ideas you have written.

   b   Cover the 'Balanced forces' differences circles, and see if you can remember what you have written. Use the 'Unbalanced forces' circles as clues to help you.

   c   Cover the 'Unbalanced forces' differences circles, and see if you can remember what you have written. Use the 'Balanced forces' circles as clues to help you.

d   Finally, without looking, ask a partner to test you to see if you can remember everything that you have written in your double bubble. If you have forgotten anything, draw a picture next to it to help you remember it next time.

e   Relook at your double bubble map a week later and see how much you can remember. By revisiting your ideas at a later date, your brain will retrieve information from your long-term memory, this helps you remember it better in the future.

Mass and weight

Do you remember?

Mass and weight

The **mass** of an object is a measure of the amount of matter it contains. Mass is measured in kilograms (kg) or grams (g).

Weight is the force experienced by a mass that is being pulled down by gravity. Weight is measured in newtons (N).

Activity 1

On Earth a mass of 1 kilogram experiences a weight of 10 newtons.

Complete the table below. One has already been done for you. Use a calculator if it helps you.

Object	Mass in kg	Weight in N
Book	0.1	1
Chair	3.5	
School bag	5.2	
Table		84
Pencil case		3
Teacher's cat	10	

Activity 2

A Stage 6 learner has completed a similar activity to the one you have just completed. Their answers are written in **bold**.

They are not sure they have fully understood the relationship (link) between the mass and the weight on Earth yet.

Object	Mass in kg	Weight in N	✗ or ✔
Apple	0.02	**0.02**	
Shoe	1	**100**	
Bag of potatoes	2	**200**	
Rucksack	20	**200**	
Skateboard	5	**500**	
Soccer ball	0.43	**4.3**	

1 Check and mark their work in the column (✗ or ✔).

2 What mistakes have they made? Explain why these are mistakes.

_____

_____

_____

_____

3 Write advice to help them not make these mistakes again in the future.

_____

_____

_____

Do you remember?

The effect of gravity on mass and weight

The mass of an object stays the same, as it is the amount of matter the object contains.

The weight of an object changes depending on what other objects are pulling down on it.

Therefore, if an object moves away from Earth to a different place in the universe, even though its mass will stay the same, its weight will change.

Activity 3

On Earth, a mass of 1 kilogram experiences a weight of 10 newtons.

On the Moon, a mass of 1 kilogram experiences a weight of 6 newtons.

On Jupiter, a mass of 1 kilogram experiences a weight of 25 newtons.

1 Use the table that you completed in Activity 1 to complete the table below.
   Add the weight for each object if they were taken to the Moon or Jupiter. One has been done for you.

Object	Mass in kg	Weight in N on Earth	Weight in N on Moon	Weight in N on Jupiter
Book	0.1	1	0.6	2.5
Chair	3.5			
School bag	5.2			
Table		84		
Pencil case		3		
Teacher's cat	10			

2 Where would be the hardest place to lift the teacher's cat up, Earth, the Moon or Jupiter? Explain your answer.

_____

3 Where would you be able to jump the highest? Explain your answer.

_____

Gases, floating and sinking

Do you remember?

Gases and mass

Gases are made from matter that is too small for us to see with our eyes. Because they are made of matter, they also have a mass.

Different gases are made from different matter, and this means that different gases can have different masses.

Activity 1

Did you know that to help scientists compare the masses of different gases, they created a unit called the molar mass? Molar mass is measured in grams per mole (g/mol), where a mole is a set number of particles in a gas, not a furry mammal!

What other scientific units do you already know? _____

The mass of dry air in the atmosphere is 29 g/mol.

1 What do you think would happen to a balloon of gas if it was released in the atmosphere? Circle the answer for each type of gas. Explain your reasoning for each answer.

a A balloon of nitrogen with a mass of 14 g/mol.

Float up   Sink to the ground   Stay where it is

The reason for my answer is _____

_____

b A balloon of oxygen with a mass of 32 g/mol.

Float up   Sink to the ground   Stay where it is

The reason for my answer is _____

_____

c A balloon of carbon dioxide with a mass of 44 g/mol.

Float up   Sink to the ground   Stay where it is

The reason for my answer is _____

_____

d   A balloon of helium with a mass of 4 g/mol.

Float up   Sink to the ground   Stay where it is

The reason for my answer is _____

_____

2   Thinking about the answers you have just given, what rule can you
come up with to help you predict whether gases will float or sink in air?

_____

_____

_____

_____

_____

_____

Do you remember?

Weight in water

All masses can apply a force on other objects. This means that solids, liquids and gases can
apply a support force on other objects they interact with.

If a liquid can push back enough on an object, it will float. If a liquid cannot push back enough
on the object, it will sink.

Activity 2

A Stage 6 learner incorrectly thinks that their surfboard floats on water and feels lighter
because there is less gravity in the water acting on the surfboard. They have drawn the following
force diagrams:

1   On a separate sheet of paper, draw your own force diagrams for the learner carrying the
surfboard in air and in water, do not forget to fully label your force arrows following the force
arrow rules.

Pull force from hand
on the surfboard

Weight of pull of
Earth on surfboard
because of gravity

A Surfboard in air

Pull force from hand
on the surfboard

Weight of pull of Earth on
surfboard because of gravity

B Surfboard in water

2   Write an explanation of how the diagrams show the scientific reason for why the surfboard floats and feels lighter in water.

_____

_____

_____

_____

_____

3   Compare your diagrams to those drawn by the learner. What idea has the learner not fully understood yet?

_____

_____

_____

Do you remember?

Upthrust

The name of the force that pushes up on objects in water is **upthrust**.

Different properties of objects affect how much a liquid can push back on them. The properties of an object that affect the force of upthrust include the mass of the object and the shape of the object.

This is how metal boats containing lots of cargo are able to float on water, even though the metal they are made from is heavier than water.

Activity 3

1   On a separate sheet of paper, use the information to draw a force diagram for each picture below.

 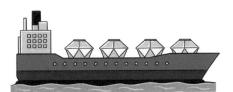

2   For each of the objects, state if the object will float or sink. Explain your reasoning.

3   For each of the objects in your answers to question 2, where you predicted that they would sink, what could you do to change the object so that it could float in the water?

4   Write a rule about floating that includes the words, **weight** and **upthrust**.

5   Write a rule for sinking that includes the words, **weight** and **upthrust**.

Revision approach

Connecting ideas in forces

Concept map

A concept map is another type of memory map. A concept map is a useful way to help you remember key words and ideas in a topic and help you check that you have understood how ideas link together.

When you draw a concept map you have to use words and you can then show how ideas link by drawing lines between them. You then need to write on the lines why you are connecting the lines together.

Activity 4

Create a forces concept map

You will need:
- plain paper cut up into small squares
- scissors
- large sheet of paper
- coloured pens

1 Cut up the plain paper into small squares.

2 Write each of the words below onto a piece of the square paper.

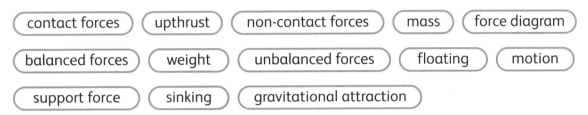

contact forces) (upthrust) (non-contact forces) (mass) (force diagram

balanced forces) (weight) (unbalanced forces) (floating) (motion

support force) (sinking) (gravitational attraction

3 Group the words in a way that makes sense to you.

4 Lay the groups of words onto the large sheet of paper, so they are separated out, and give titles to the groups you have made.

5 Now try to make connections between different words, add lines between them and write on the line why you have connected the words. For example, you might want to draw a line between the words 'air resistance' and 'upthrust' and you could label the line 'both are contact forces that push up on objects'. Look at an example of a concept map below.

6 When you have finished making as many connections as you can, take a photo of your completed concept map. You can use this photograph to help you revise the key words and how ideas connect in this topic.

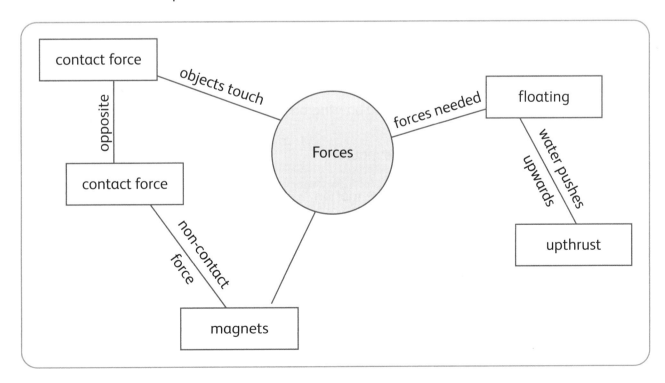

Revision quiz

Revisit the work that you have done in this unit, check your key word cards, read through them to remember scientific words and information. Use the concept map that you made in Activity 4 on page 65 to remind you about forces. When you are ready, complete this short quiz.

As you work through it you can help yourself by:

● Reading each question carefully – check you understand the question.

● Look for key words, use them in your answer.

● Answer the question in your mind first before you write it.

● If the question is multiple choice decide which answers you know are definitely incorrect, then think about the answers that are left, to decide which one is correct.

● Check your answer to make sure that you do not want to make any changes.

Answer the questions on a separate sheet of paper or in your notebook.

1 Forces are caused when:

   a one object pushes on another

   b two objects interact with each other

   c one object pulls on another

2 Forces that act on objects that are touching are called:

   a Contact forces

   b Non-contact forces

   c Gravity

3 a What unit do we measure force in?

   b What instrument do we use to measure force?

   c What are the instruments measuring:

     distance

     mass

     weight

   d What do the readings on the scale tell you about the apple and the bowling ball?

   e What is the mass of the apple and bowling ball?

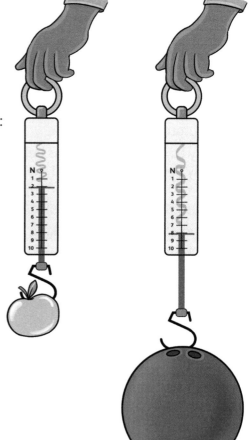

4 Is the statement true or false?

   a Balanced forces can change the motion of an object.

   b Unbalanced forces can change the motion of an object.

5 Match the words to the correct sentences.

   a Steady speed A motion that happens when forces acting on an object are unbalanced.

   b Stationary   The motion that happens when balanced forces act on a moving object.

   c Speeding up The motion that happens when balanced forces act on a still object.

6   Write four rules we should use when drawing force arrows on force diagrams.

7   Name three different contact forces.

8   Use the words **weight** or **mass** to complete the sentences. The first one is done for you.

The a _____ of an object tells us how much matter there is in the object.

This is not a force. b _____ is not related (linked) to how the objects interact.

c _____ is the force experienced by an object that is being pulled down by gravity.

The d _____ of an object will stay the same no matter where it goes in the universe.

The e _____ of an object will change depending on what different object interacts with it.

The f _____ of an object changes depending on where in the universe it is.

9   On a piece of graph paper, draw a graph of the mass of the object against its weight on Earth using the information in the table.

Object	Mass in kg	Weight in N
Apple	0.02	0.2
Shoe	1	10
Bag of potatoes	2	20
Rucksack	20	200
Skateboard	5	50
Soccer ball	0.43	4.3

a   Write a title for the graph.

b   What should the label be on the *x* axis?

c   What should the label be on the *y* axis?

d   Plot the different items in the table on the graph.

e   Mark on the graph what the mass would be for an object with a weight of 80 N.

f   Mark on the graph what the weight would be for an object with a mass of 8.5 kg.

10 True or false?

a   Gases have a mass.

b   Gases that are lighter than the atmosphere sink.

c   Gases that are heavier than the atmosphere sink.

d   Objects that weigh less than the force of upthrust in water float.

e   Objects that weigh more than the force of upthrust in water float.

Unit 6 — Electrical circuits

What will you learn

This unit will help you to revise your learning of:
- electrical circuits can be drawn as a circuit diagram using symbols
- circuits can be parallel or series
- series and parallel circuits can be compared to observe the effect on components such as lamps.

🔗 Pages 81–92

Electrical circuits

Revision approach

Electrical circuit key word cards

You have been making key word cards for different topics, and you might already have made cards for the standard circuit symbols used throughout the world (see Learner's Book page 83). For this topic, you need to make additional cards for the key words you need to learn as well as the circuit symbols.

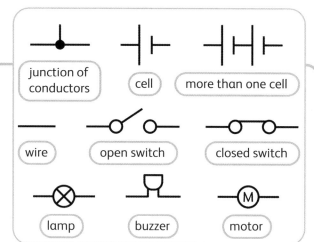

junction of conductors cell more than one cell

wire open switch closed switch

lamp buzzer motor

Activity 1

Here is a list of the words that you need to learn and remember about electrical circuits.

You will need:
- small pieces of card or plain paper
- coloured pens

(mains electric sockets) (cells)

(electrical conductivity) (component) (circuit) (circuit symbols) (parallel circuit)

(junction of conductors) (cell) (more than one cell) (wire) (open switch)

(closed switch) (lamp) (buzzer) (motor) (circuit diagram) (series circuit)

1   Write each word on the front of a card, add any images that help you remember what the word means.

2   On the back of your card write a definition. Check your definition using a dictionary. Think about whether your definition is a good one or if parts of it need changing. Add as many other ideas that you can think of that link to the key word on the front.

3   Are there any words you do not know or find hard to write a definition for? Spend time finding out more about those words; use the Learner's Book, if you have one, to help you with this. Ask a partner to check what you have written.

4   Work with a partner who has also made these cards and play a game of 'Snap'. Together, at the same time, turn over one card to show the front picture and place it in front of you. Keep doing this until you both turn over the same picture. The first person to correctly shout out the word on the card wins, and they collect all of the cards that have been turned over and add them to their pack. The winner is the person who can collect all of the cards from their partner.

Do you remember?

Circuit symbols and circuit diagrams

Electrical conductivity is how well a material conducts electricity. A complete electrical loop (with no gaps) is needed for any electrical component to work. We call this complete electrical loop a **circuit**.

Electrical circuits will not work if there are breaks anywhere in the loop. This means that for circuits to work:

- there needs to be a source of electrical power (cell or mains)
- all the components must be connected correctly
- there must be no gaps in the circuit.

A **circuit diagram** shows the components in a circuit and how they are connected to one another. Circuit diagrams use standard circuit symbols.

Activity 2

The standard circuit symbols do not always look like the objects they are representing. One of the skills of becoming a scientist is being able to draw the correct scientific circuit diagram for any circuit.

You will need:
- pencil
- paper
- ruler

1 Practise your scientific circuit drawing skills by drawing all of the pictures below into the correct circuit diagram.

2 Ask a partner to check the circuit diagrams you have drawn.

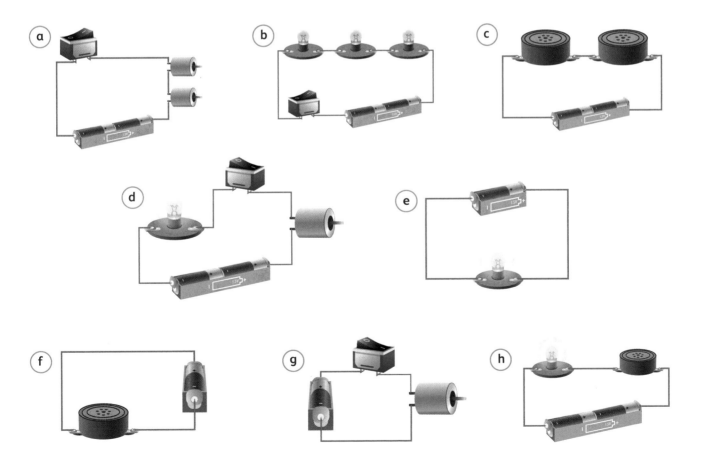

Activity 3

Several Stage 6 learners have been asked to draw the circuit diagram for the circuit on the right.

Look carefully at the circuit diagrams produced by different Stage 6 learners to represent the picture.

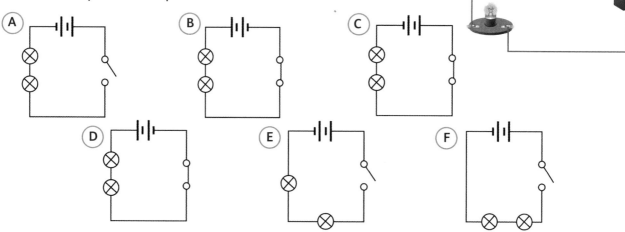

1 Which learner/s have drawn a correct diagram to represent the circuit? _____

2 Who has made mistakes in their drawing? Explain where they have made errors.

_____

_____

3 Write some feedback for the learners who made mistakes telling them what they need to change to improve their diagram.

_____

_____

Science in context

Alessandro Volta

Alessandro Volta was born into a noble family in Italy in 1745. Volta's parents were worried because he did not start talking until he was four years old. The delay in speaking did not seem to harm him. As he got older, not only could he speak Italian, he also learnt Latin, French, German, English, Spanish, Dutch, Russian and Ancient Greek.

Volta became interested in doing experiments and from a young age he wrote to important scientists in France. In 1775, when he was working at the Royal School in Como (Italy), he developed his first invention.

He continued to carry out experiments and question ideas, and this led him to making a discovery about how electrical batteries work. He said that what was needed to start off the electrical flow was the contact between two different metals.

Other scientists disagreed with his ideas and it divided the international scientific community. Eventually, everyone agreed with Volta because of his most famous invention: the Voltaic Pile or electric battery.

Designed in 1799, the Voltaic Pile was made of alternating copper and zinc discs with a weak acidic layer separating each pair of metals. The contact between the metals produced electricity in the form of sparks.

This battery became the first continuous source of electric current. The battery quickly became known around the world and the French Emperor, Napoleon Bonaparte, asked the inventor for a demonstration of it. Amazed by the device, the emperor named Volta Count and Senator of the Kingdom of Lombardy.

Volta died at his home in Italy in 1827 at the age of eighty-two. In his honour, Volta has a crater on the Moon and an asteroid named after him.

Activity 4

Draw a cartoon strip about Alessandro Volta.

You will need:
- pencil and coloured pens
- large sheet of paper

- To make your cartoon strip, fold your paper in half, fold it in half again and fold it in half once more. When you open your paper, you should have 8 sections in it.
- Read the information about Volta and decide what the eight most important facts are.
- The eight facts you have chosen will be the different pictures you will draw in the 8 sections you have on your sheet of paper.
- In your cartoon strip you need to use at least 10 key words, one of which must be the word 'battery'.
- When you have finished making your cartoon strip, read the story to a partner. To help to remember information about Alessandro Volta, you can use the cartoon strip you have drawn.
- Were you able to tell the whole story and use all of your key words? If there was anything you forgot on your cartoon strip, add more details and key words to your drawing to help you.

After you have told the story to a partner using only your cartoon strip, try and answer the questions below.

1 Which country was Volta born in? _____

2 Why were Volta's parents worried about him when he was young? _____

_____

3 What was Volta's most famous invention? _____

4 How did Volta develop his science ideas? _____

_____

5 Did other scientists welcome his ideas? Explain your answer. _____

_____

_____

6 What do you think is the most important part of the story and why? _____

_____

_____

Series and parallel circuits

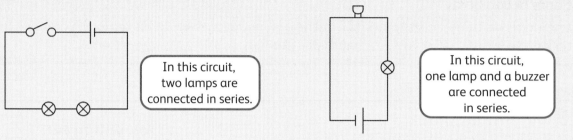

Do you remember?

Series circuits

In a circuit diagram, if there is only one electrical loop that can be traced with your finger, this is called a **series circuit**. In a series circuit, all the components are connected in the same loop.

In this circuit, two lamps are connected in series.

In this circuit, one lamp and a buzzer are connected in series.

Activity 1

Changes in series circuits

When the number of components in a series circuit changes, this will have an effect on the different components in the circuit.

For each of the different components below, describe what could be observed if the number of components in the series circuit was increased. One example has been done for you.

Component	Observable change
Buzzer	Buzzer volume will decrease
Lamp	
Motor	

Activity 2

Predicting changes in series circuits

Once you understand how changes in a series circuit affect the different components in it, you can use this to make predictions.

1   Copy and complete the table below by looking at the circuit in the first column and predicting what you think will happen for each situation. Explain your reasons, one has been done for you.

Circuit	Situation	Prediction	Reasons
	Another lamp is added to the circuit.	I think … Both bulbs will be dimmer.	Because … The two bulbs only have one cell now lighting both of them.
	Another cell is added to the circuit.	I think …	Because …

Circuit	Situation	Prediction	Reasons
	A lamp is removed from the circuit.	I think …	Because …
	A cell is removed from the circuit.	I think …	Because …
	The motor is removed from the circuit.	I think …	Because …
	Another motor is added to the circuit.	I think …	Because …
	The buzzer is removed from the circuit.	I think …	Because …
	Another cell is added to the circuit.	I think …	Because …

2   Having completed the table, write three golden rules to help other Stage 6 learners make predictions about circuits. What would your Golden Rules be?

i   _____

ii   _____

iii   _____

Do you remember?

Parallel circuits

If components are connected in more than one loop, this is called a **parallel circuit**.

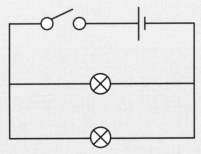 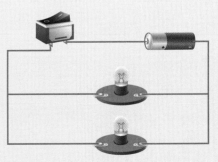

Even though they look different, both of these electrical circuits have two lamps in parallel. This is because a complete electrical circuit has to include the cell and a component, in this circuit the components are lamps.

Activity 3

You can check the number of loops in a circuit by putting your finger on the cell and then moving it along the wire via components until you reach the other side of the cell. If you trace the loops in the picture on the previous page with your finger, you can see that each lamp is in a separate loop.

1   For the circuit diagrams below, decide:

    a   The number of loops in each circuit.

      A _____ B _____ C _____

      D _____ E _____ F _____

    b   Which circuits are the same.

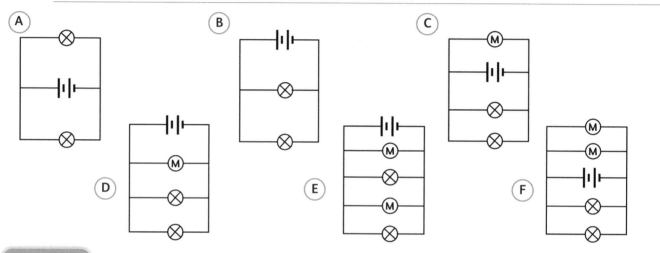

Activity 4

Predicting changes in parallel circuits

A Stage 6 learner found that if they changed how the components in a circuit were connected, it changed the way the components worked.

The learner investigated the circuits on the right.

The learner created a table to record their results and they have filled in the first example.

1   Copy and complete the table below by looking at the circuit in the first column and predicting what you think will happen for each situation. Explain your reasons, one has been done for you.

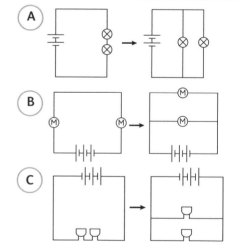

Circuit	Prediction	Reasons
A	I think … Both bulbs will be brighter	Because … Each loop in the parallel circuit now only has one bulb in it instead of two when it was in a series circuit.
B	I think …	Because …
C	I think …	Because …

2 Having completed the table, write another Golden Rule to help other Stage 6 learners make predictions about circuits. What would your new Golden Rule be?

_____

Activity 5

Handling data

A class of Stage 6 learners investigated fruit batteries. The class set up their investigation like the picture below:

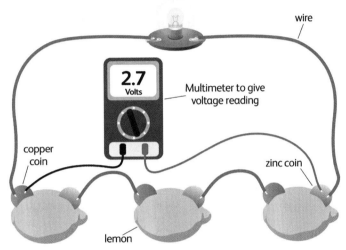

The class looked at how increasing the number of lemons changes the voltage made.

1 What is the dependent variable in their investigation?

_____

2 What is the independent variable in their investigation?

_____

3 What would the learners have to control in their investigation to make the results collected from the experiment valid? Results that are valid help a scientist answer the original question they were investigating.

_____

_____

The data collected by the learners is in the table below.

Number of lemons	Voltage reading on multimeter in volts (V)
1	0.9
2	1.8
3	2.7
4	3.6
5	4.5
6	
7	6.3

4   On a piece of graph paper, draw a line graph to show the data in the table on the previous page. Remember to give your graph a title and to label both axes.

5   The learners forgot to collect their data for six lemons, use your graph to predict what reading they should get on the multimeter. Explain your reasons.

6   The learners wanted to run LED lights using their lemon battery. The learners had an LED light that needed 2.2 V and another LED light that needed 4.2 V. How many lemons would the learners therefore need to make:

a   just the 2.2 V LED light work? Explain your reasons.

b   just the 4.2 V LED light work? Explain your reasons.

c   both the 2.2 V LED and the 4.2 V LED light work in the same circuit? Explain your reasons.

Activity 6

Creating a series and parallel circuits double bubble

You used a double bubble as a memory map in Unit 5 Forces. Revisit page 58 if you need a reminder.

You will need:
- plain paper
- coloured pens

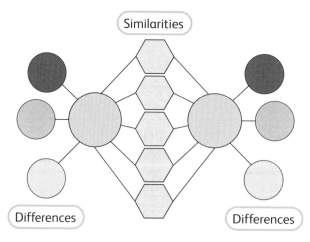

Similarities

Differences   Differences

1   Draw a double bubble memory map like the example above, or use the double bubble template from your teacher.

2   In the large left circle write 'Series circuits' and in the large right circle write 'Parallel circuits'.

3   Write as many ideas as you can in the middle hexagons that are the same for both 'Series circuits' and 'Parallel circuits'. For example, *they need wires*. Add extra hexagons if you need to.

4   In the outside circles, make a note of ideas that are different in 'Series circuits' and 'Parallel circuits'. For example, in the purple circles you could write for 'Series circuits', *only one loop*; and for 'Parallel circuits', *more than one loop*. Add extra circles if you need to. Try and make the coloured circle ideas connect. In the example above the person is thinking about the number of loops.

5   Ask a partner to check your double bubble to see if you have got the ideas correct.

6   To test your memory:

   a   Cover the middle hexagons and see if you can remember all of the ideas you have written.

   b   Cover the 'Series circuits' differences circles, and see if you can remember what you have written. Use the 'Parallel circuits' circles as clues to help you.

   c   Cover the 'Parallel circuits' differences circles, and see if you can remember what you have written. Use the 'Series circuits' circles as clues to help you.

   d   Finally, ask a partner to test you to see if you can remember everything that you have written in your double bubble. If you have forgotten anything, draw a picture next to it to help you remember it next time.

   e   Revisit your double bubble map a week later and see how much you can remember. By revisiting your ideas at a later date, you retrieve information from your long-term memory, this helps you remember it better in the future.

Revision approach

Learning flower

A learning flower is a memory aid that can help you remember and organise ideas. This will help you recall important things you have covered in this topic and make links between the different ideas.

Activity 7

Create a learning flower to connect ideas in electrical circuits

1   Cut out a circle for the centre of your flower, petals to go around the edge, a stem, leaves and some roots.

2   In the centre circle of your flower, write the topic title.

3   Make petals for all of the key words you need to remember in this topic, look back at the key words from the start of the topic. On the front of each petal, write the key word and on the back of each petal write a definition for your word. Use a dictionary or glossary if you need help.

4   On the stem, write what you think are the most important things you have learnt from this unit.

5   On the roots, write the things that you already knew about this topic.

6   On the leaves, write any questions that you have about this unit, or ideas that you feel you do not yet fully understand.

7   Once you have created all the different parts of your flower, glue it down on your large sheet of paper. Glue the petals down using the tip that touches the circle so that you can read both sides.

8   Spend time either talking to a partner about what you have written on your leaves, or spending time doing some further research to help you find the answers.

You will need:
- coloured paper
- scissors
- large sheet of paper
- coloured pens
- glue stick

Revision quiz

Go back over the work that you have done in this unit, check your key word cards, read through them to remember scientific words and information. Use your learning flower to remind you about electrical circuits. When you are ready complete this short quiz.

● Read each question carefully – check you understand the question.

● Look for key words, use them in your answer.

● Answer the question in your mind first before you write it down.

● Look carefully at the circuit diagrams, especially the number of cells in a circuit.

Answer the questions on a separate sheet of paper or in your notebook.

1 Which of the following are components?

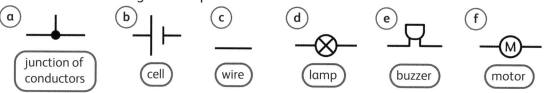

2 Series circuits:

    a have more than one loop

    b have components connected in different loops

    c make the brightness of lamps decrease if more are added

3 Parallel circuits:

    a have more than one loop

    b make the brightness of lamps decrease if more are added in a new loop

    c have components connected in the same loop

4 True or false?

    a Electrical conductivity is how well a material conducts electricity.

    b Electrical circuits must be complete for components to work.

    c All materials conduct electricity.

5 Match the words to the correct sentences.

    a Switch   An electrical circuit that has only one loop.

    b Series circuit   An electrical circuit that has more than one loop.

    c Parallel circuit   A component used to close and open gaps in an electrical circuit.

6 Draw the circuit symbols for the following:

    a cell

    b open switch

    c closed switch

    d wire

7 Name the three different components pictured below.

    a   **b**   **c**

8   Decide if the answer to these questions is **series**, **parallel** or **both**.
    a   Which circuit has only one loop in the electrical circuit?
    b   Which circuit has more than one loop in the electrical circuit?
    c   Which circuit has components that are in different loops?
    d   Which circuit has components that are all in the same loop?
    e   Which circuit changes how the components work if more cells are added?
    f   Which circuit changes the brightness of lamps if more lamps are added?

9   Look at the two circuits.

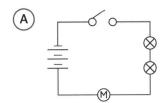   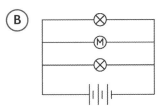

    a   In which circuit will the lamps still be on if the motor breaks?
    b   In which circuit will the lamps be the brightest?
    c   In which circuit will the motor be spinning the fastest?

10 True or false?
    a   Traffic light lamps should be wired as a series circuit.
    b   A string of lights should be wired as a series circuit.
    c   Lights in a house should be wired as a series circuit.
    d   If you want to switch off all the lights in a house with a single
        switch, the switch needs to be in series to the lights.

What will you learn

This unit will help you to revise your learning of:
- light travels in straight lines
- when a ray of light is reflected from a plane mirror it changes direction
- when a ray of light travels through different mediums (for example, glass and water) it changes speed and can change direction, this is called refraction
- reflected light ray bounces off the material at the same angle as the incoming light ray.

 Pages 93–103

Reflection

Revision approach

Light key word hexagons

To help you remember the key words of this topic you are going to make key word cards. However, for this topic your key word cards will be in the shape of a hexagon.

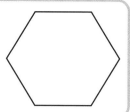

Activity 1

Here is a list of the words that you need to learn and remember about light.

(light rays) (reflect) (light detectors) (light sources) (opaque)

(transparent) (translucent) (reflections) (angles) (diffuse)

(specular) (mirror) (mirror image) (angle of incidence)

(angle of reflection) (periscope) (boundary) (refraction)

You will need:
- coloured card
- tracing paper
- scissors
- coloured pens
- hexagon cards
- hexagon template

1   Write each word on the front of a card, add any images that will help you remember what the word means.

2   On the back of your card write a definition. Check your definition using a dictionary. Think about whether your definition is a good one or if parts of it need changing. Add as many other ideas as you can that link to the key words on the front.

3   Shuffle the cards, place them in a pile and pick up the top card.

4   On the hexagon template, write a question where the answer would be the word on the card that you have picked. Do the same for the next card on the top of the pile until you have created a question for every card.

5   When you have written questions for your cards, ask a partner to help you.

6   Layout your hexagons with the words showing.

7   Ask your partner to read out the questions you have written on your hexagon template, you then need to find the card that is the answer and place the card over the question on the hexagon template. You could time how long it takes you to find the answers to all the questions. You can then try and do it again later in this topic, and again at the end to see if you can beat your previous time.

Do you remember?

Opaque objects and light – reflection

All opaque objects reflect light, even black ones. However, some opaque objects allow you to see the image of the object in them. These clear images that can be seen are called **reflections**.

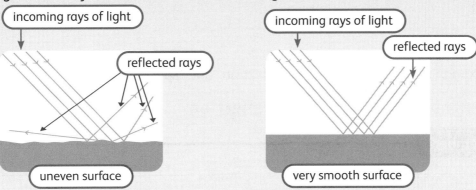

When rays of light shine on most opaque surfaces, the rays are reflected at lots of different angles, because the surfaces are uneven and rough. Rough surfaces do not usually make clear images. This is called **diffuse reflection**.

When rays of light hit a very smooth opaque surface, all the rays reflect at the same angle. This produces reflections and is called **specular reflection**.

Activity 2

Look at each of the pictures below.

Ⓐ Ⓑ Ⓒ Ⓓ Ⓔ Ⓕ Ⓖ Ⓗ Ⓘ Ⓙ

1 Sort the pictures into two groups, one group for examples that show diffuse reflection and one group for examples that show specular reflection. Any that you are unsure of, leave to the side and do some research to see if you can find out the answer.

Diffuse reflection	Specular reflection

2　What features do your diffuse reflection examples have in common?

3　What features do your specular reflection examples have in common?

4　Create hexagon cards for both 'diffuse reflection' and 'specular reflection'.

Angles of incidence and angles of reflection

Do you remember?

The diagram shows a light ray reflecting from a mirror.

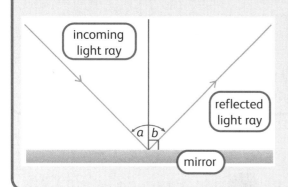

A 'Normal' line is drawn from where the light hits the mirror at 90° to the surface (a 'Normal' line is drawn at right angles to the surface).

Angle a is the angle between the Normal and the incoming light ray (we can tell by the direction of the arrow). This is called the **angle of incidence**. Angle b is the angle between the Normal and the reflected light ray (we can tell by the direction of the arrow). This is called the **angle of reflection**.

Activity 1

A group of Stage 6 learners conducted an investigation to find out if there was a pattern in their results that would answer their question:

Is there is a relationship (link) between the angle of incidence and the angle of reflection?

The learners shone a light ray onto a mirror and measured the angle of incidence and the angle of reflection.

The learners then repeated the activity using different angles of incidence and recorded their results. These can be seen in the table below:

Experiment	Angle of incidence (in degrees)	Angle of reflection (in degrees)
1	15	15
2	25	25
3	38	40
4	52	
5	65	65
6	76	74
7		81

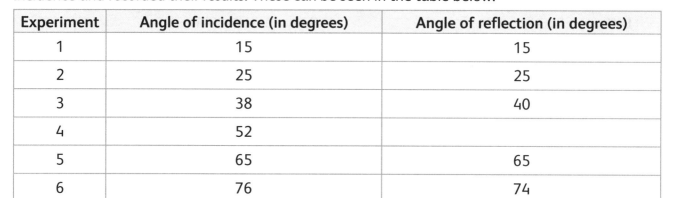

1 What type of science enquiry were the learners carrying out? Circle your answer from the choices below.

   Research   Fair test   Observe over time   Identify and classify   Pattern seeking

2 Explain the reasons for your choice.

   _____

   _____

3 The learners forgot to make a note of two of their results and they are missing from the table. What values do you think they would have got? Explain the reasons for your values.

   Row 4: _____

   _____

   Row 7: _____

   _____

4 Where do you think there are errors in the results the learners collected? Why do you think this might have happened? What should the learners have done to improve the accuracy of their results?

   _____

   _____

   _____

   _____

5 Draw a diagram of a protractor like the one below. Use a pencil and a ruler and carefully add the lines for all of the experiments 1–7. A line has already been drawn for Experiment 5. Label each of your lines with the experiment number and make sure to add the arrows to show the direction of the light ray.

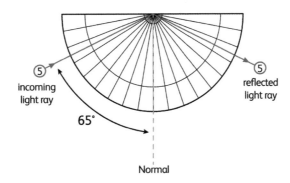

6 Use the pattern in the results to draw a conclusion to answer the question, '*Is there is a relationship (link) between the angle of incidence and the angle of reflection?*'

   _____

   _____

Do you remember?

Everyday applications of reflection

There are lots of different ways people use mirrors in everyday life.

- Rear-view mirror in a car – allows a person to see traffic coming from behind.
- Dental mirror – allows a dentist easier access to see the upper and back teeth.
- Baby toy with a small mirror – allows the baby to see his or her own reflection.
- Security mirror (such as in a shop) – allows security personnel to observe customers and their activities.
- Mirror wall in a dance studio – allows dancers to see and check their posture and dance movements.
- Kaleidoscope – a cylinder with mirrors inside it. It contains loose objects such as small pieces of coloured plastic or glass. You look into one end of the kaleidoscope and hold the other end up to the light. The mirrors and objects inside it create colourful patterns, made up of many reflections.

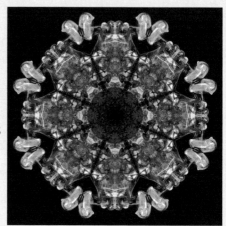

Activity 2

A group of Stage 6 learners investigated how the angle of mirrors affected the number of pencils that were seen in a kaleidoscope they made. The results the learners collected can be seen in the table below.

Angle of mirrors	Number of pencils
90°	4
60°	6
45°	8
36°	10
30°	12
20°	18

1 What was the question the learners were investigating?

_____

2 Which variable is the independent and which is the dependent in the learners' investigation?

Independent variable: _____

Dependent variable: _____

3 Use the data in the table to describe the pattern between the angle of the mirrors and the number of pencils you see.

4 The learners predicted how many pencils they would see if the mirrors were placed at an angle of 120° to each other. They said there would be 2. Do you agree or disagree with them? Explain your reasons.

5 What rule can you come up with to describe the relationship between the angle of the mirrors and the number of pencils seen? Write your rule. If possible, compare it to a partner's, are they the same or different?

6 Write three more ways mirrors are used in everyday life and explain how they help people.

Refraction

Do you remember?

Transparent objects and light – refraction

When light meets a transparent object, it is not reflected. It travels through the object.

The speed of the light ray changes at the boundary (edge) of two different materials. For example, it changes speed when the light ray goes from air into glass, or from water into air.

Because the speed changes, this often means that the direction of the ray changes. We call this change in speed and direction, **refraction**.

Refraction can make objects appear differently when we look at them.

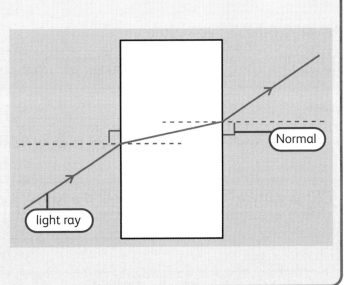

Stage 6 learners were learning that light travels at different speeds through different transparent objects. The objects they researched were:

The learners did some research to find the different speeds of light and wrote their results in table below.

Material	Speed of light in kilometres per second (km/s)
Water	225 000 km/s
Diamond	124 000 km/s
Acrylic (plastic)	200 000 km/s
Air	Approx. 300 000 km/s
Glass	200 000 km/s

1 Rank the results of the speed of light in order from the fastest to the slowest.

_____

2 What patterns do you notice about the type of material and the speed of light?

_____

_____

Having ordered the speeds of light, learners wrote a conclusion.

Learner 1 – The speed of light changes in different materials. I think it changes because the materials are different.

Learner 2 – The speed of light changes when it travels in different transparent materials. The speed of light is fastest when it travels through air, which is a gas. The speed of light then slows down when it travels through water, which is a liquid. The speed of light then slows even more when it travels through different solids. The speed of light is slowest of all when it travels through diamond. I think the type of material changes the speed of light because of the arrangement of particles it is made of. Gases slow it the least, then liquids and finally solids. The closer together the particles are, the more the speed of light slows down.

Learner 3 – The speed of light changes when it travels in different transparent materials. The speed of light is slowest when it travels through air, which is a gas. The speed of light then speeds up when it travels through water, which is a liquid. The speed of light then speeds up again when it travels through different solids. The speed of light is fastest of all when it travels through diamond. I think the type of material changes the speed of light because of the arrangement of particles it is made of. The further apart the particles are, the more the speed of light slows down.

Learner 4 – The speed of light is different in different transparent materials. As the materials change so does the speed. The speed of light is fastest in air and slowest in diamond. I think the transparent material changes the speed.

3   Read the conclusion written by each of the learners and rank them in order from the best explanation to the weakest explanation.

_____

4   Explain your reasons for the order you have put them in.

_____

_____

_____

_____

5   What speed do you think the speed of light will be in sea water? Explain your reasons.

_____

_____

6   A haiku is a poem made of 17 syllables written in three lines. There are five syllables in the first line, seven in the second and five in the last. Here is an example:

Lit up by the light

of the bright cold winter moon

silent owl takes flight

Write your own haiku to help you remember the order of the speed of light in different transparent materials.

_____

_____

_____

> **Do you remember?**
>
> **Law of refraction**
>
> When light goes from one transparent material to another, its speed can change:
> - Light that travels from a solid to a liquid, or from a liquid to a gas, will get faster.
> - Light that travels from a gas to a liquid, or from a liquid to a solid, will get slower.

Activity 2

Stage 6 learners performed a role play to explain how the direction of the light ray changed when it travelled from one transparent material to another.

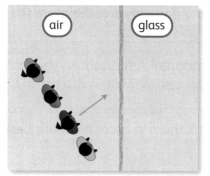

The learners discovered that:
- As the light travels from a solid to a liquid, or from a liquid to a gas and speeds up, it bends away from the Normal.
- As the light travels from a gas to a liquid, or from a liquid to a solid, it slows down and bends towards the Normal.

Learners then drew ray diagrams to show what was happening as light travelled across the boundaries of different materials.

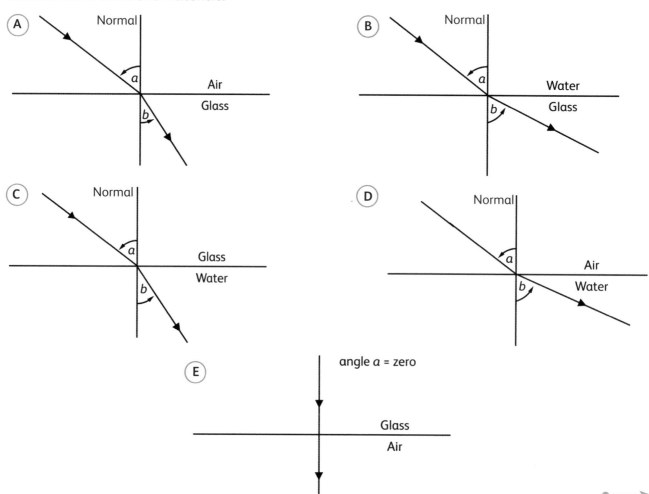

1   Look at the diagrams the learners drew and sort them into a True and False grid, giving the reasons for your decision.

True	False

2   For any examples that you have decided are false, redraw them on a separate sheet of paper, so that they are correct.

3   On a separate sheet of paper, draw some diagrams of your own. You may find it useful to look back at the speed of light table on page 86. Your examples need to show light rays travelling from:

a   Plastic to air

b   Diamond to water

c   Water to plastic

d   Water to air

e   Plastic to diamond

f   Diamond to plastic

Do you remember?

Applications using the law of reflection

Because the speed of a light ray changes when travelling from one transparent material to another, objects in or behind the transparent materials can look different to how they actually are.

Scientists have developed many different types of lenses to help us in our everyday lives. There are many different devices that have lenses in them. These include:

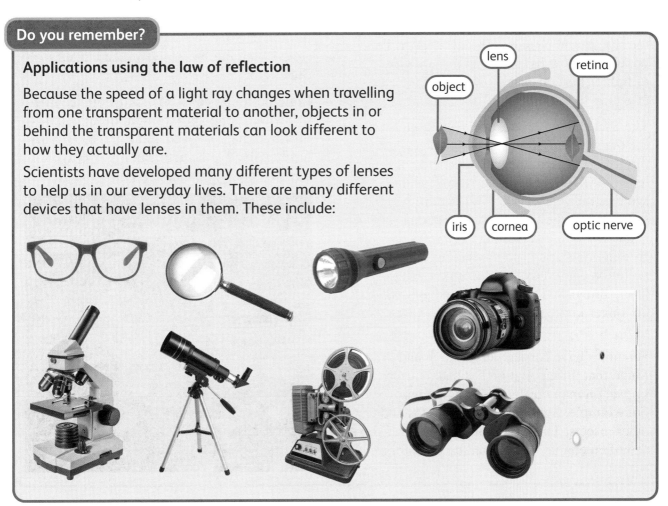

Activity 3

Images are powerful ways of helping our brain remember things.

Including our eye there are nine different objects that use lenses shown on the previous page.

1 Look at the pictures for 10 seconds.

2 Cover all of the pictures and see if you can write out all nine objects without looking in 10 seconds.

3 If there were any you could not remember, have a look again at the pictures, add them to the list you made by drawing a picture.

4 Give your list of nine objects to someone else and get them to test you to see if you can name all of the nine different objects.

5 Follow instructions 1–4 again two days later, did you remember all nine objects? Keep practising this every two days until you can name all nine without forgetting any.

Science in context

Impact of light pollution

Light pollution happens when the night sky is brightened by street lamps and other human made sources of light such as advertising boards, floodlights, skyscrapers, office blocks and our homes. This light pollution means that during the night, when the sky should be dark, 80% of the world's population now lives with a constant glow of light. This constant glow of light is affecting living things.

The Sun is like a clock, and even though the time of sunrise and sunset change throughout a year, the pattern of the length of daylight across a year stays the same as the year before. This pattern is so reliable (regular) that it gives plants and animals signals for natural cycles of feeding, migrating and navigating. The light pollution caused by humans is confusing living things and many plants and animals no longer know if it is day or night.

Scientists have carried out research and found that light pollution has had a negative impact on many different living things. For example, light bulbs attract moths and other insects. The insects at these bulbs can often end up being eaten by predators because they are more easily seen. If insects are attracted to the headlights in moving vehicles, they can be killed from being hit by the vehicle. All of this has resulted in a huge decline in insect populations. Birds also fly into lit buildings and can die, scientists have estimated that between 100 million and one billion birds die every year from flying into buildings in the United States alone.

Light pollution can also change the feeding behaviour of nocturnal (night time) animals, making it harder for them to find food. Bats have evolved to be active at night to avoid daytime predators that use vision to hunt,

such as birds of prey, and light pollution is causing bats to leave their roosts and avoid areas that are too well lit, impacting on the local ecosystem.

Some insects such as fireflies, rely on bioluminescence (the ability of an organism to give off light) to help them find and attract a mate. Light pollution can confuse males and make it difficult for them to find females, and also difficult for them to see one another's signals.

Light pollution can disrupt the nesting behaviour of turtles. On a natural, dark beach, the horizon above the sea is slightly brighter than the horizon above land. Hatchlings (baby turtles) can detect this difference and this helps them find the quickest path to the sea. Light pollution can confuse the hatchlings and instead of crawling to the sea, they either head towards the lights or crawl in circles and get lost on the beach. Hatchlings do not have the ability to think that they are doing something wrong; they simply crawl towards the brightest light, and this often leads them to death.

Humans are also badly affected by the increased use of light bulbs. Light from these bulbs can affect our sleep patterns, making it harder for the brain to 'switch off' and fall asleep, as well as making it more difficult to stay asleep, all of this reduces the quality of sleep for people. Exposure to artificial light at night has also been linked to different illnesses such as diabetes, some cancers and can affect a person's moods.

There are a number of different ways we can reduce the impact of light pollution. These include turning off lights at night and reducing the number of windows that are lit during the hours of darkness. Researchers estimate that dimming a building in this way could reduce bird deaths by 60%. Many countries have now passed laws looking at ways to reduce light pollution including requiring outdoor lighting to be shaded and keep light below set levels of brightness. Each of us can help, by doing things such as closing curtains at night (so the light does not escape), only lighting what needs to be lit and focusing security lights carefully.

Activity 4

Read the article about the impact of light pollution.

You will need:
- a sheet of paper
- coloured pens
- fishbone organiser template

1   List all of the ways that light pollution is created on Earth – these are your causes.

_____

_____

_____

2   List all of the impacts that light pollution can have on different living things – these are your sub-effects.

_____

_____

_____

_____

3   List all of the ways that humans can try and reduce the levels of light pollution – these are your solutions.

4   A fishbone organiser is a way to link many ideas that relate to the same issue. An example of a fishbone organiser is shown below:

Cause and effect fishbone organiser

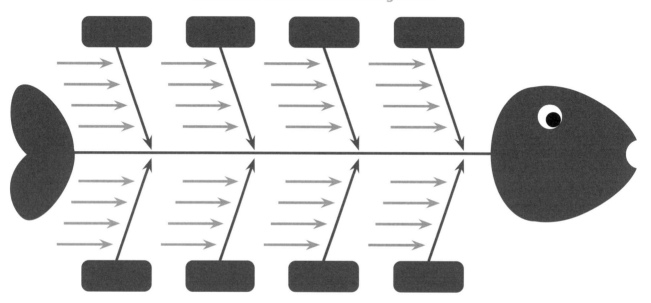

Each of the bones in the diagram will list the causes at the top with sub-effects and solutions linked to the cause.

a   Draw a fishbone organiser or use the template from your teacher and write the title 'Light pollution' in the head. This is the main effect you are learning about.

b   At the top of each bone, write one of your causes.

c   For each of your causes, add in any of the sub-effects caused by it on the ribs above the spine, and any solutions that could be used to reduce it on the ribs below the spine.

d   Now use your fishbone to learn all of the different causes, sub-effects and solutions linked to light pollution.

e   Give your fishbone to a partner and ask them to test you to see if you can name and link all of the different parts related to the complex issues of light pollution.

f   If you have forgotten anything, draw a picture next to it to help you remember it next time.

g   Revisit your fishbone organiser a week later and see how much you can remember. By revisiting your ideas at a later date, you retrieve information from your long-term memory, which helps you remember it better in the future.

Revision approach

Light

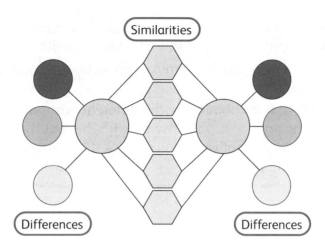

1   Draw out a double bubble memory map or use the double bubble template from your teacher.

2   In the large left circle write 'Reflection' and in the large right circle write 'Refraction'.

3   Write as many ideas as you can in the middle hexagons that are the same for both 'Reflection' and 'Refraction'. For example, *they affect light rays*. Add extra hexagons if you need to.

4   In the outside circles make a note of ideas that are different in 'Reflections' and 'Refraction'. For example, in the purple circles you could write for 'Reflections', *happens when light meets an opaque object;* and for 'Refraction', *happens when light meets a transparent object.* Add extra circles if you need to. Try and make the coloured circle ideas connect. In the example above, the person is thinking about what happens when light meets different objects.

5   Ask a partner to check your double bubble to see if you have got the ideas correct.

6   To test your memory:

    a   Cover the middle hexagons of your double bubble memory map and see if you can remember all of the ideas you have written.

    b   Cover the 'Reflections' differences circles and see if you can remember what you have written. Use the 'Refraction' circles as clues to help you.

    c   Cover over the 'Refraction' differences circles and see if you can remember what you have written. Use the 'Reflections' circles as clues to help you.

    d   Finally, without looking, ask a partner to test you to see if you can remember everything that you have written in your double bubble. If you have forgotten anything, draw a picture next to it to help you remember it next time.

7   Revisit your double bubble a week later and see how much you can remember. By revisiting your ideas at a later date, you retrieve information from your long-term memory, this helps you remember it better in the future.

Revision approach

Hexagons

Now you have finished revising your learning for this topic, you are going to use your key word hexagon cards that you made at the start of this unit, see page 80.

Hexagons are a useful way to help your brain make connections between different ideas that you have been learning about.

Place your hexagons together, think about how the different hexagons that touch each other are linked. There are no right or wrong answers with this, but there are right or wrong explanations. There are many different ways you can link the hexagons, the key is that you can explain the links between them.

Activity 5

Create a hexagon map to connect ideas in light

1   Use the hexagon cards that you made at the start of this unit, see page 80. Feel free to make new hexagons for any additional words you would like to add to your hexagon map.

You will need:

- hexagon cards

2   Layout and place hexagons with sides touching.

3   Say how the hexagon cards that you have placed are connected to each other. For example, you could put the hexagon cards for 'reflect' and 'opaque' next to each other and say that 'all objects that are opaque reflect light'. You could then add in the hexagon card for 'diffuse' so that all three cards are touching each other. You could then say that 'opaque objects that have rough surfaces cause light to reflect at lots of different angles and this is called diffuse reflection'.

4   See if you can place all of your hexagons onto your hexagon map.

5   To challenge yourself you can:

- Shuffle the cards and stack them up. Take each card from the top of the pile in the order they are stacked and place them as a hexagon map, talking about the links every time you place a card down.

- Pick out seven cards, place one in the centre and place the other six cards around it and explain links between the touching cards.

You can take photos of your hexagon map if it helps you remember the connections you have made.

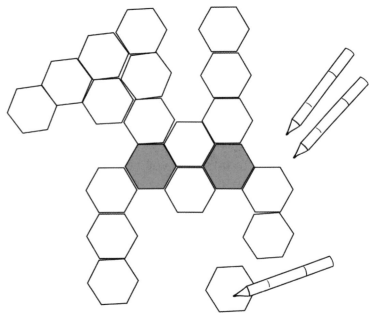

Revision quiz

Revisit the work that you have done in this unit, check your hexagon key word cards, read through them you to remember scientific words and information. Use your hexagon learning map to remind you about light. When you are ready, complete this short quiz.

As you work through it you can help yourself by:

- Reading each question carefully – check you understand the question.
- Look for key words, use them in your answer.
- Answer the question in your mind first before you write it.
- If the question is multiple choice decide which answers you know are definitely incorrect, then think about the answers that are left, to decide which one is correct.
- Check your answer to make sure that you do not want to make any changes.

Answer the questions on a separate sheet of paper or in your notebook.

1   Which of the following are examples of specular reflection?

aluminium can	tree	glass of water	balloon
mirror	sponge	metal spoon	pen
ice cubes	clothes	soil	silver ring

2   Reflection:
    a   occurs when light meets transparent objects
    b   occurs when light meets opaque objects
    c   occurs only from shiny objects

3   Refraction:
    a   occurs when light meets transparent objects
    b   occurs when light meets opaque objects
    c   occurs when light travels through air

4   True or false?
    a   When a ray of light is reflected, the angle of incidence is the same as the angle of reflection.
    b   When a ray of light is refracted, the angle of incidence is the same as the angle of refraction.
    c   All opaque materials reflect light.

5   Match the words to the correct sentences.
    a   Angle of incidence   A line drawn at 90° where light hits the surface.
    b   Angle of reflection   The angle between the Normal and the incoming light ray.
    c   Angle of refraction   The angle between the Normal and the reflected light ray.
    d   Normal   The angle between the Normal and the refracted light ray.

6   On a sheet of paper, accurately draw the ray diagrams for the following:

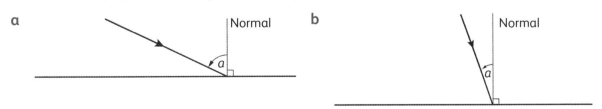

7 Name five different everyday uses of lenses.

8 Decide if the answer to these questions is **reflection**, **refraction** or **both**.

a What can occur when a ray of light meets an object?

b This involves an incoming ray of light.

c What occurs when light meets an opaque object?

d What occurs when light meets a transparent object?

e What occurs when the angle of the incoming and outgoing ray of light are usually different?

f What occurs when the incoming and outgoing rays of light always have the same sized angles?

g What occurs for the speed of light to change?

9 Look at the diagram, you can use any of the following words in your answer.

(air) (water) (glass) (ice)

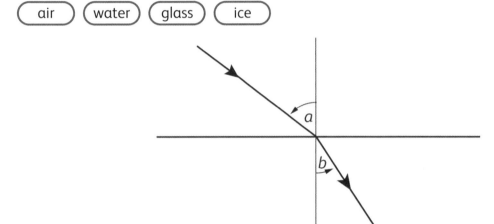

a Give one example of what material 1 and material 2 could be, explain your reasons.

b Give another example of what material 1 and material 2 could be, explain your reasons.

10 True or false?

a Light slows down when it meets a transparent material.

b Light speeds up when it meets a transparent material.

c Light slows down when it goes from a gas to a liquid.

d Light speeds up when it goes from a liquid to a gas.

e Light speeds up when it goes from a liquid to a solid.

11 Artificial light pollution is causing many problems on Earth.

a Name three causes of the artificial light pollution.

b Name three problems artificial light pollution causes.

c Name three ways humans can help reduce the problem of artificial light pollution.

What will you learn

This unit will help you to revise your learning of:

- the Earth is made up of four main layers called the crust, the mantle, the outer core, and the inner core
- rocks can be classified as metamorphic, igneous, and sedimentary
- the rock cycle explains how one type of rock can be changed into another over time
- weathering and erosion are part of the rock cycle
- fossils are found in sedimentary rocks, there are cast fossils and trace fossils
- there are different types of soils, we can classify them as sand, clay and silt and loam. Soils can be changed.

 Pages 104–125

The layers of the Earth

Revision approach

Rocks key word concept map

Look at the list of words on the next page, which words do you already know? Create a concept map for the words that you can already remember. Use small pieces of card, and as before, add the word and an image on the front of your card by either drawing pictures or cutting and sticking pictures out of magazines.

Do not worry if you do not know all of the words yet. You can make cards for any words you do not know when you meet them as you revise this unit. You will then do this activity again at the end of the topic, and you should then be able to make more connections.

When concept map word cards are placed on the concept map you will draw lines between them. You then need to write on the line your reason for linking the concept map word cards. For example, you could draw a line between the words **granite** and **hard**, and write on the line, '*granite is an example of a hard rock*'.

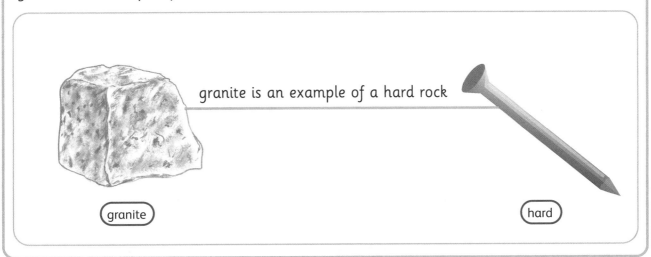

granite is an example of a hard rock

granite · hard

Activity 1

Create a concept map

By the end of this unit, you should be able to know and remember the words about rocks below.

You will need:
- card
- scissors
- coloured pens
- large sheet of plain paper

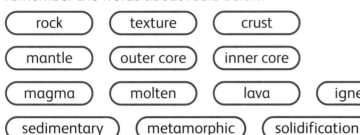

(rock)   (texture)   (crust)

(mantle)   (outer core)   (inner core)

(magma)   (molten)   (lava)   (igneous)

(sedimentary)   (metamorphic)   (solidification)   (granite)   (basalt)

(limestone)   (chalk)   (sandstone)   (marble)   (particles)   (erosion)

(sediments)   (sedimentation)   (burial)   (metamorphism)   (fossil)

(cast fossil)   (trace fossil)   (geologists)   (palaeontologists)   (index fossils)

(fossil record)   (rock cycle)   (weathering)   (soil)   (organic)   (humus)

(topsoil)   (subsoil)   (bedrock)   (sand)   (inorganic)   (silt)

(clay)   (drainage)   (loam)   (soil erosion)

1   Make concept map word cards for the words that you recognise. Write each word on the front of the card, and add any images that help you remember what the word means.

2   Layout the concept map word cards on the plain paper.

3   Now draw lines to join as many of the cards as you can on the sheet of paper. On the connecting lines, write the reason for the link.

4   The more lines you make, the more links you are able to show.

5   When you have finished, take a photograph of your concept map, and collect and keep your concept map word cards safe, you will need them again at the end of this unit.

Do you remember?

The Earth's layers

The Earth is made up of a number of different layers. These layers are called the crust, the mantle, the outer core, and the inner core.

The Earth's crust is solid and made from much cooler rocks than the mantle. The material in the mantle behaves like a very thick liquid that flows, a little like toothpaste does when you squeeze the tube gently. The outer core is the liquid part of the centre of the Earth, and the inner core is the solid part of the centre of the Earth.

Although the rocks at the surface can feel cool or cold, miners and cavers can feel an increase in temperature as they go down into the Earth's crust.

(crust)
(mantle)
(outer core)
(inner core)

Activity 2

Stage 6 learners were asked to bring in an object to model the different layers of the Earth.
The learners brought in the following:

A   An Indian scotch egg.

B   An apple sliced in half.

C   A peach sliced in half.

D   A wrapped chocolate sweet that has a liquid centre with a nut in the middle.

E   A kiwi sliced in half.

1   Complete the table stating how the different parts of each object model the different layers of the Earth; one has been done for you.

Object	Inner core	Outer core	Mantle	Crust
Indian scotch egg	Egg yolk	Egg white	Meat	Breadcrumbs
Sliced apple				
Sliced peach				
Wrapped sweet				
Sliced kiwi fruit				

2 Rank the models in the order from the one you think is the best to the one you think is the weakest.

_____

3 Explain the reasons behind the order you ranked the models.

_____

_____

4 Think of another object the learners could have used to model the Earth's layers.

_____

5 Make up a song or rap to help you remember the layers of the Earth and whether the layer is a solid or a liquid.

6 Make up four actions to use when singing your song, one action each for crust, mantle, outer core and inner core.

7 Perform your song to a partner and teach them your actions.

Do you remember?

Volcanoes, magma and lava

The mantle layer of the Earth is very hot and thick, and is made of rocky material and magma. Magma is made of molten (melted) rock. When this molten rock (magma) escapes to the Earth's surface, it is called lava.

Volcanoes are found in the crust of the Earth. Volcanoes allow hot lava, volcanic ash, and gases to escape from magma chambers below the surface. On Earth most volcanoes are found underwater, these are called submarine volcanoes.

Activity 3

Stage 6 learners used a wax, sand and water model to represent how submarine volcanoes are formed. Learners were asked to produce a cartoon strip of how the volcano was formed.

Here is one of the learner's drawings.

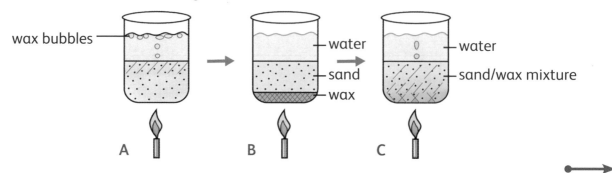

However, the learner's work has been hack attacked. Someone has removed all of the key labels and put the pictures in a different order.

1 What order should the pictures be in to show how submarine volcanoes are formed?

_____

2 On a separate sheet of paper, copy the drawings in the correct order and label them, where appropriate, using all of these science key words:

Earth's crust   **lava**   **sea water**   **magma chamber**   **volcano**

3 Do you think this is a good model for how submarine volcanoes are formed?

a What do you think are the good parts of the model?

_____

_____

_____

b What do you think are some of the weaknesses of the model?

_____

_____

_____

4 One learner suggested that the model shows how islands are formed in the ocean. What evidence does the model provide to back up the statement that the learner made?

_____

_____

_____

Types of rocks

Do you remember?

Igneous rocks

There are three types of rocks on Earth: **igneous** rocks, **sedimentary** rocks and **metamorphic** rocks.

These rocks are formed in different ways. Each type is able to change into another type.

basalt   granite

Igneous rocks form when magma from a volcano cools and turns into a solid. This process is called **solidification**. As the hot rock cools down to form igneous rocks, crystals form inside the rocks.

There are two kinds of igneous rocks:
- extrusive igneous rock, which form outside the Earth's crust, like basalt
- intrusive igneous rock, which form within the Earth's crust, like granite.

Activity 1

A group of Stage 6 learners conducted some research to learn about igneous rocks. They produced a map of their work.

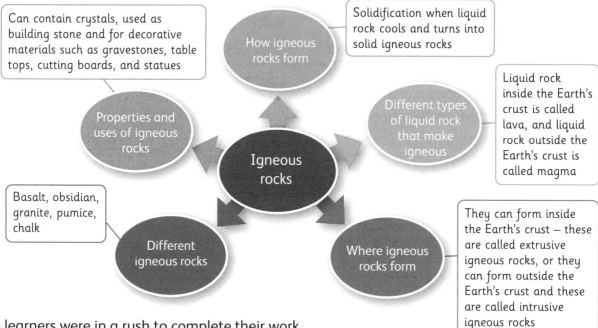

Can contain crystals, used as building stone and for decorative materials such as gravestones, table tops, cutting boards, and statues

Properties and uses of igneous rocks

How igneous rocks form

Solidification when liquid rock cools and turns into solid igneous rocks

Different types of liquid rock that make igneous

Liquid rock inside the Earth's crust is called lava, and liquid rock outside the Earth's crust is called magma

Igneous rocks

Basalt, obsidian, granite, pumice, chalk

Different igneous rocks

Where igneous rocks form

They can form inside the Earth's crust – these are called extrusive igneous rocks, or they can form outside the Earth's crust and these are called intrusive igneous rocks

The learners were in a rush to complete their work.

1   Check their map to see if everything they have written is correct.

2   If you spot any errors they have made, write below what they should have written instead.

Activity 2

The learners then created an acrostic using the word **IGNEOUS** to help them remember how igneous rocks are formed. Their acrostic is written below.

You will need:
- plain paper
- coloured pens

    Inside

ma**G**ma

 ca**N**

    Escape

    Outside

    Undergoing

    Solidification

1   On a separate sheet of paper, draw a picture that the learners could hold up to show what their acrostic is describing.

2   Look back at the key word list for this topic on page 98. Review the last few pages and create concept map word cards for any words you have now come across that you did not make cards for earlier.

Do you remember?

Sedimentary rocks

Sedimentary rocks are formed from **particles** (small pieces) of materials over thousands or even millions of years.

Rocks break down over time into **sediments** (small rock particles), and **erosion** (the movement of broken-down rock from one place to another) then happens. Erosion happens by wind, water and ice, moving the sediment to rivers, lakes or oceans.

The sediment then settles at the bottom of the water, a process known as **sedimentation**. Over time, all of the sediment creates layers, like layers of sand.

Burial of the sediment then happens as the thickness of the layers increases over time.

Over millions of years, the particles of rock become squashed together and form firm rock.

Activity 3

1 Produce a cartoon strip to explain how sedimentary rocks are formed. You need:

  - at least 6 cartoon sections in your explanation
  - to use at least 8 key words, one of which must be 'sediments'
  - to show how sedimentary rocks link to igneous rocks.

2 Give your cartoon strip to a partner and get them to test you to see if you can describe each of the boxes in the correct order. Explain how one box follows on from another and use all of the key words correctly.

You will need:
- coloured pens
- large sheet of plain paper

Do you remember?

Erosion and sedimentation

Weathering and erosion are processes that break down rocks and the difference between them depends on where the process takes place.

Weathering is a process that breaks up rocks where they are located. Weathering happens by rain, wind, water, plants and animals.

When the smaller pieces of broken-down rock are moved from one place to another, this is called **erosion**. Erosion happens by wind, water and ice, and the broken-down rock is moved to rivers, lakes or oceans.

Weathering often leads to erosion by causing rocks to break down into smaller pieces, which erosive forces can then move away.

So, if a rock is changed or broken but stays where it is, it is called weathering. If the pieces of weathered rock are moved away, it is called erosion.

Activity 4

Stage 6 learners wanted to answer the question: *Which size of rock particles are the most easily eroded?*

The learners modelled the process of erosion and sedimentation by using a piece of guttering filled with a sand mixture, they then poured water at the top end of the guttering.

The learners used magnifying glasses to examine the sand mixture at the top of the guttering before the water was poured. They then examined the sand mixture at the top of the guttering again after the water was poured.

The learners also observed what happened to the water as it collected in the water tank over time.

The learners refilled the guttering with the sand mixture and repeated their experiment.

The information that the learners collected is in the table below.

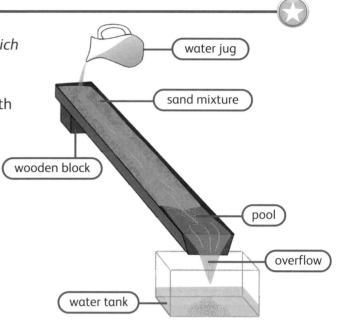

Description of sizes of rocks	Experiment 1	Experiment 2
At top of guttering before pouring water	Mixture of different sized rocks (lots of large pieces bigger than 1 cm, lots of small pieces smaller than 0.5 cm).	Mixture of different sized rocks (lots of large pieces bigger than 1 m, lots of small pieces smaller than 0.5 cm).
At top of guttering after pouring water	Mostly larger pieces of rocks bigger than 1 cm, only a few smaller rocks of less than 0.5 cm left.	Mostly larger pieces of rocks bigger than 1 cm, only a few smaller rocks of less than 0.5 cm left.
In water tank after pouring water	Water was cloudy and took 5 mins 20 seconds to turn clear. Mostly smaller pieces of rocks less than 0.5 cm found at bottom of tank.	Water was cloudy and took 6 mins 27 seconds to turn clear. Mostly smaller pieces of rocks less than 0.5 cm found at bottom of tank.

1   What type of science enquiry were the learners carrying out? Circle the correct answer from the choices below.

Research   Fair test   Observe over time   Identify and classify   Pattern seeking

2   Explain the reasons for your choice.

_____

_____

3   Why did the learners decide to do the experiment again?

_____

_____

4   Were the observations the learners collected able to help them answer the question they were investigating? Explain your answer.

_____

_____

5   What did the observations the learners collected show about which size of rock particles are the most easily eroded?

_____

_____

_____

6   Why do you think these are the rock sizes that are most easily eroded?

_____

_____

Do you remember?

Metamorphic rocks
Metamorphic rocks are formed from igneous and sedimentary rocks that have been heated or squashed in the Earth's crust. The rise in temperature and pressure cause the rocks to change. This process of change is called **metamorphism**. The rocks are heated but are not hot enough to melt and turn into magma.

Activity 5

Stage 6 learners modelled how igneous, sedimentary and metamorphic rocks were created and formed from each other.

They did this by using dark, milk and white chocolate. Below are their instructions.

A
Carefully remove your three solid chocolate rocks from the aluminium foil pie dishes and grate half of each of them.   In your new aluminium foil pie dish, build up different layers of the different colours of grated chocolate.   When you have built up layers to fill the pie dish, carefully apply pressure by pressing down on the top of the chocolate with your spoon.
B
Add any solid chocolate rock you have left to the top of the squashed layers of chocolate rock in the aluminium foil pie dish. Carefully place the pie dish so that it floats on top of the hot water.

Observe as the water heats the chocolate and the chocolate starts to melt. Carefully remove the aluminium foil pie dish from the bowl when the chocolate is soft to the touch. Use your spoon to gently test for when this happens.

Before the chocolate cools down, gently squash and fold the mixture with your spoon.

C

Break up one type of chocolate into chunks and put the chunks inside a piece of aluminium foil.

Now do the same for the other types of chocolate you are using. Carefully place each aluminium foil case, containing chocolate, into a bowl, so that each one floats on top of the hot water.

Observe and time how long it takes for the water to heat and melt the chocolate until a smooth liquid forms. Carefully remove the aluminium foil with your molten chocolate from the bowl and leave it to cool.

1   Which instruction A, B or C is for making igneous rocks? Explain your reasons.

_____

_____

2   Which instruction A, B or C is for making sedimentary rocks? Explain your reasons.

_____

_____

3   Which instruction A, B or C is for making metamorphic rocks? Explain your reasons.

_____

_____

4   Which order would you put the instructions in? Explain your reasons.

_____

_____

Activity 6

Flipbooks

You will need:
- coloured pens
- sheets of plain paper
- stapler or sticky tape

1   Create a flipbook using pictures and key words, to show how igneous, sedimentary and metamorphic rocks are made, and how they are then produced from each other.

2   The key words you need to include in your flipbook are:

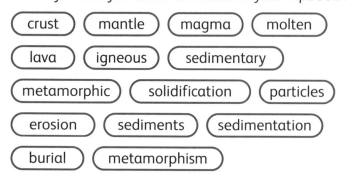

(crust) (mantle) (magma) (molten)

(lava) (igneous) (sedimentary)

(metamorphic) (solidification) (particles)

(erosion) (sediments) (sedimentation)

(burial) (metamorphism)

3 Revisit the key word list for this topic on page 98. Review the last few pages and create concept map word cards for any words you have now come across that you did not make cards for earlier. Add them to your concept map, do not forget to use lines to make links between the words and write the reason for each link.

Fossils and the rock cycle

Do you remember?

Fossils

Fossils are the dead remains of parts of plants and animals from millions of years ago. The plant or animal was covered quickly after death by sediments such as mud and sand. In time, the sediments hardened to form rock and the bodies inside the sediments formed fossils, called **cast fossils**.

Some fossils are not formed by the actual plant or animal body but by things left behind by the animals or plants, such as eggs, solid wastes (their droppings) and tracks. These fossils are known as **trace fossils**.

Activity 1

The Stage 6 learners did an activity where they sequenced pictures and matched sentences to show how fossils have been formed over millions of years.

These were the pictures and sentences they were given.

A   The hard parts of the animals remained. They were squashed and covered by new rock.

B   More bits of rock that were washed down by rivers covered the rotting animals.

C   After death, mud or sand covered some animals.

D   After a very long time, the land changed and the animal remains changed to rock. We call these fossils.

Below are the answers that three different learners gave:

Learner 1

I think the order of the pictures and sentences is:

1D 2A 3C 4B

Learner 2

I think the order of the pictures and sentences is:

1A 3B 2D 4C

Learner 3

I think the order of the pictures and sentences is:

3A 2B 1C 4D

1   Which learner has got the sentences and pictures in the correct order? Explain your reasons.

_____

_____

2   Is the fossil in the picture a trace fossil or a cast fossil? Explain your reasons.

_____

_____

3   A learner says that fossils are only formed in igneous rocks. Is their statement true or false? What is your evidence for your answer?

_____

_____

4   If you have not already made concept map key word cards for fossil, cast fossil and trace fossil, make them now and add them to your pile.

Do you remember?

Rock cycle

In science, we use cycles to help us explain a regularly repeated sequence of events. Scientists realised that rocks of one kind can change into rocks of another kind. They called this the **rock cycle**.

Activity 2

Drawing from memory

You will need:
- A4 sheet of plain paper
- coloured pens or pencils

Drawing from memory is a useful revision strategy to help you remember key information related to an image. Engaging more of the visual parts of your brain helps you create a richer experience to help you absorb ideas and remember them better.

The image on the next page shows a poster that Stage 6 learners put together to represent the rock cycle. The image shows the Earth's crust and magma beneath the surface, there is a volcano, a mountain and the ocean.

The learners have added 11 labels of key words and arrows to show where the different types of rocks are formed and how they are changed into each other.

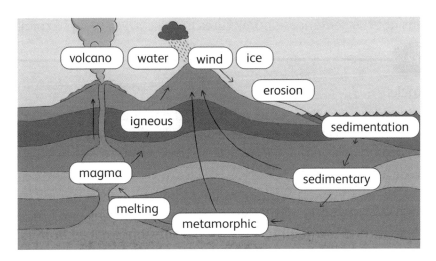

Thirty second challenge!

1   Look at the image for 10 seconds and then cover it.

2   See how much of the image you can redraw on your sheet of paper and how many of the 11 labels you can add. You can take as long as you want, but you are not allowed to look at the image yet! When you have finished, turn over your drawing so you cannot see it.

3   When you are ready, uncover the image and look at it for another 10 seconds. What did you miss? Try and remember and then cover the image again.

4   Go back to your drawing and add the things that you missed the first time. Again, you can take as long as you need, but do not be tempted to look at the image! Once you have added all you can remember turn over your drawing so you cannot see it.

5   You now have your last 10 seconds to look at the image. Try and remember the things that you have not got on your drawing yet. When your time is up, cover over the image.

6   This is the final chance for you to complete your drawing and include all of the details the learners put in their poster. Take as long as you need.

7   Now uncover the image and put your drawing next to it. Are there any things that you did not get on your drawing? Add these now and put a circle around them to help you remember them.

8   There are some key words that the learners have missed out, for example, lava. Look at the concept map key word cards that you have been making. If any key words are missing, add them to the most suitable place on your picture.

9   Give your drawing to a partner and ask them to:

   a   Test you on all of the key words that you have put on your diagram. Can you name them all?

   b   Ask you to describe where the three different types of rocks are found in the rock cycle.

   c   Pick any of the key words and ask you to describe the full rock cycle and how rocks of one type can change into another.

10 Put your drawing away and come back to it in a few days' time. Ask a partner to test you again to describe everything in the rock cycle. If you forget anything, add a circle around it, and add a picture next to it to help your brain remember.

Soil

Do you remember?

Soil made from rocks

Rocks are broken down by rain, wind, water, plants and animals. This process is called weathering. Weathering of rocks can form **soil**. Soil can be black, red, yellow, green, brown and even purple! The type and colour of soil depends on the rocks that made it.

Soil is divided into different layers.

At the surface is the darkest layer. It is made mainly from **organic** materials (the remains of dead plants and animals). Bacteria in the soil feed on these remains and break them down. As the remains of plants and animals rot, they form a substance called **humus**.

The thicker layer under the surface is a mixture of humus and rocky fragments and is called the **topsoil**. It is the part of the soil that farmers and gardeners use for growing their plants.

Below the topsoil is the **subsoil** (*sub* means *under* or *below*). This layer is paler than the topsoil because it contains much less humus. Below the subsoil is a layer containing lumps of rock, this layer is called **bedrock**.

Activity 1

1   For each of the statements below decide if the answer should be **weathering** or **erosion**. If you cannot remember what erosion is, go back and look at page 103.

   a   This happens to a mountain when a tree grows its roots into it. _____

   b   This happens when a glacier moves down a valley. _____

   c   If a gravestone is looked at after 50 years this will have caused it to change. _____

   d   This happens when animals that live underground cause the rocks around them to weaken

       and breakdown. _____

   e   This is the reason the ocean floor is constantly changing. _____

   f   This happens when rocks in deserts get very hot during the day and very cold at night. The outer rock layer weakens and breaks down the rock over a long period of time, like an onion skin.

       _____

   g   If water gets inside cracks in a rock and the temperature overnight drops below freezing point, the water turns into ice. Over time this can break down the rock.

       _____

   h   This happens when people walking through the countryside cause the paths to wear away.

       _____

2   Create hand signals to represent the words **weathering** and **erosion**, to help you remember these words and how they are different.

Activity 2

The Stage 6 learners made an edible model of soil.

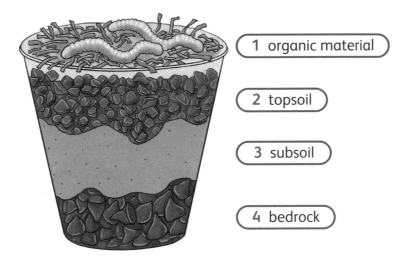

1 organic material

2 topsoil

3 subsoil

4 bedrock

The learners used the following edible material to build the different layers of their model.

crushed biscuits   chocolate and butterscotch chips   shredded coconut coloured green
gummy worms   chocolate mousse

1   Which materials did the learners use to represent the bedrock layer? Explain your reasons.

_____

_____

_____

2   Which materials did the learners use to represent the subsoil layer? Explain your reasons.

_____

_____

_____

3   Which materials did the learners use to represent the topsoil layer? Explain your reasons.

_____

_____

_____

4   Which materials did the learners use to represent the organic material layer? Explain
    your reasons.

_____

_____

_____

Do you remember?

Different types of soil

There are three main types of soil: **sand**, **silt** and **clay**. The main difference between sand, silt and clay is the particle size of the rocks they are made from.

- Sand is made up of larger particles that we can see with the naked eye. Sandy soils feel gritty.
- Silt is made up of particles too small to see with the naked eye. Silty soils feel silky.
- Clay is made up of tiny particles that fit together very closely. Clay soils feel sticky. Clay soils are not as easy to squeeze together as the other types of soil.

The materials that soils are made of come from both inorganic sources, such as sand, silt and clay and organic sources like the remains of dead plants and animals.

Loam is a type of soil that is a mixture of 40 % sand, 40 % silt and 20 % clay. It also has large amounts of organic materials, called **humus**, which is formed by the decomposition of leaves, other plant material and animal matter by soil microorganisms.

Activity 3

You are going to make fact files about the four different types of soil: sand, silt, clay and loam.

You can use the information given above and do some of your own research.

You will need:
- pieces of card
- pen or pencil

1. Your fact file needs to include the following:
   - A title
   - Information about the size of rock particles in your soil type
   - Details of how the soil is formed
   - Where the soil can be found
   - A drawing of what the soil can be used for.

2. When you have finished all four fact files, give them to a partner. Ask them to ask you anything from any of the fact files and see if you can answer all their questions correctly.

Changing soil and the Wood Wide Web

Soil is one of the Earth's most important natural resources and 25% of all species on Earth live in it!

Scientists have discovered that within the soil there is an underground network of plant roots and fungi which are working together. Scientists have called this underground fungal network the 'Wood Wide Web'. Scientists believe the Wood Wide Web allows trees to share nutrients between each other, and that it may also trap carbon dioxide in the soil.

If the Wood Wide Web does take carbon from the atmosphere and trap it in the soil, then fungi could help the planet. We know that carbon is a big contributor to climate change and locking carbon away could help to slow down climate change. However, scientists do not know very much about the fungal networks, and want to find out more to see if they can be used to help fight climate change.

Scientists have therefore, decided to carry out an investigation. Scientists are going to collect 10 000 samples from the soil to create a world map of where the fungal networks are located. They will then use computers to help them work out what the function of these fungal networks are and how they could help slow down climate change.

Current estimates put the amount of carbon dioxide taken out of the air and locked up in the soil with the help of fungal networks at five billion tonnes – although it could be more than three times higher than this. Once they have analysed the evidence about how the Wood Wide Web is working and

how it is helping, scientists can help us think about how we can protect and look after our land and soil.

Scientists say that soil and the Wood Wide Web fungal networks are under threat due to:

- Increased use of land for agricultural – this happens when forests and grasslands are changed to farmed fields and pastures, and is called deforestation.

- Increased use of fertilisers and pesticides – these are used to kill pests and add nutrients to the soil, however, they can also kill off the fungi and damage the fungal networks.

- Increased use of land for urbanisation – this happens when green areas in towns and cities are taken and used as areas for buildings; for example, homes, factories.

Scientists estimate the total length of the Wood Wide Web fungal network in the top 10 cm of soil on the planet is more than 450 quadrillion kilometres, that is around half the width of our galaxy!

1 Read the extract above about fungi in the Wood Wide Web and create an infographic (showing information using pictures, charts or diagrams).

2 Use your infographic to help you answer the following questions:

   a What is the Wood Wide Web?

   _____

   _____

   b What do scientists think the Wood Wide Web does?

   _____

   _____

You will need:
- piece of paper
- coloured pens or pencils

c Why do scientists think that the fungal network formed by fungi in the soil is important for the planet?

_____

_____

d List three different ways that humans are damaging the Wood Wide Web. For each one, write what people could do to prevent this damage.

_____

_____

_____

Revision approach

Concept map

You have finished revising your learning for this topic and you are going to create a final concept map of the topic. Concept mapping is a useful way to help your brain make connections between different words that you have been learning about.

When concept map word cards are placed on the concept map, group them and then draw lines between them. Write on the line your reason for linking the concept map word cards. For example, you could draw a line between the words **igneous** and **fossil**, and write on the line, '*fossils cannot be found in igneous rocks as they form when liquid rock cools and all fossils would be destroyed*'.

There are lots of different ways that you can link the concept map word cards, the key is that you can explain the reasons why you think the connected cards link.

Activity 5

Create a concept map to connect ideas in rocks and soils

1 Use the concept map word cards that you made at the start, and throughout this unit.

2 Revisit page 98 to see if you now have cards for all of the key words for the unit. You can make new cards for any additional words you would like to add to your concept map.

3 Layout out the concept map word cards and sort them into groups. Decide on titles for the groups you have made.

4 Place the groups of concept map word cards down on the plain paper, when you are happy you can glue your cards down and write your titles for the groups above them.

5 Now draw lines to join as many of the cards as you can on the sheet.

6 You can join cards within a group and across different groups. The more lines you make, the more links you are able to show. Compare this concept map to the one you did at the start of the unit, how many new connections can you now make?

7 Give your concept map to a partner and ask them to:
   a Pick two words that you have connected. Explain your connections to your partner.
   b Read out one of your connections and then tell them what the two connected words are. If you cannot remember anything, add a picture to your concept map to help you remember it next time.

8 You can take photos of your concept map once you have finished it and use this to revise and help you remember the connections you have made.

Revision quiz

Go back over the work that you have done in this unit, check your concept map word cards, read through them to remember scientific words and information. Use your concept map to remind you about rocks and soil. When you are ready complete this short quiz.

As you work through it you can help yourself by:

- Reading each question carefully – check you understand the question.
- Look for key words, use them in your answer.
- Answer the question in your mind first before you write it.
- If the question is multiple choice decide which answers you know are definitely incorrect, then think about the answers that are left, to decide which one is correct.
- Check your answer to make sure that you do not want to make any changes

Answer the questions on a separate sheet of paper or in your notebook.

1 Which of the following are examples of igneous rocks?

   granite   marble   obsidian   sandstone   limestone   chalk   basalt
   slate   pumice

2 Magma:
   a   is liquid rock above the Earth's crust
   b   is liquid rock beneath the Earth's crust
   c   is solid rock above the Earth's crust

3 Lava:
   a   is liquid rock above the Earth's crust
   b   is liquid rock beneath the Earth's crust
   c   is solid rock above the Earth's crust

4 True or false?
   a   The mantle is the outer most layer of the Earth.
   b   The outer core is the inner most layer of the Earth.
   c   The outer core is made up of liquid material.
   d   Volcanoes formed under water are called submarine volcanoes.
   e   Igneous rocks formed outside the Earth's crust are called extrusive.
   f   Igneous rocks formed inside the Earth's crust are called extrusive.
   g   Igneous rocks contain fossils.

5 Match the words to the correct sentences.
   a   Igneous rocks   Liquid rock beneath the Earth's crust.
   b   Sedimentary rocks   Formed from other rocks that have been heated and squashed in the Earth's crust.
   c   Metamorphic rocks   Formed when magma from a volcano cools and turns into a solid.
   d   Magma   Formed from particles of materials over thousands or millions of years.

6   Write definitions for each of the following words:

   a   Solidification

   b   Sedimentation

   c   Burial

   d   Metamorphism

7   Name five different ways weathering and erosion can occur.

8   Decide if the answer to these questions is **weathering** or **erosion**:

   a   When a tree grows its roots into a mountain.

   b   When a glacier moves down a valley.

   c   When animals that live underground cause the rocks around them to weaken and break down.

   d   This is the reason the ocean floor is constantly changing.

   e   When rocks in deserts get very hot during the day and very cold at night and the rock breaks down over a long period of time.

   f   If water gets inside cracks in a rock and the temperature overnight drops below freezing point, the water turns into ice. Over time this can break down the rock.

   g   When people walking through the countryside cause the paths to wear away.

9   Use the diagram to help you answer the questions.

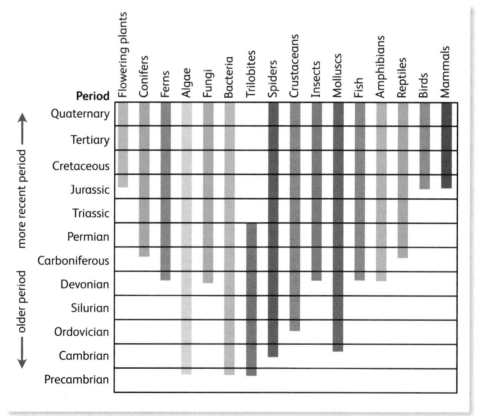

   a   Which appeared in the fossil record first – algae or fungi?

   b   How many new groups were added to the fossil record in the Carboniferous Period?

   c   Which is the group of animals that still exists and has been around the longest?

   d   In the period during which the flowering plants appeared, which animal groups also appeared in the fossil record? Why do you think this is?

10 True or false?

    a   Bedrock is the top layer of soil.

    b   Topsoil is the top layer of soil.

    c   Organic material is the top layer of soil.

    d   Sandy soil is made from the smallest rock particles.

    e   Clay soil is made from the smallest rock particles.

    f   Silty soil is made from the medium rock particles.

    g   Clay soil is the best at holding water.

    h   Sandy soil is the best at holding water.

11 List three ways humans are damaging the soil and changing its composition (what it is made of).

Earth and the Solar System

What will you learn

This unit will help you to revise your learning of:
- the Moon changes in appearance over its monthly cycle
- the solar system is made up of the planets (including Earth), moons, asteroids, comets and meteoroids that orbit around the Sun.

Pages 126–137

Earth and the solar system

Revision approach

Rich picture poster

A rich picture is a way of showing an idea, information or a process by using pictures, diagrams and individual words, phrases and colour coding. Using a rich picture can sometimes be easier to show what you know than, for example, writing sentences or paragraphs, especially if you are someone who learns and remembers pictures more easily.

Activity 1

For this unit you will be making a rich picture poster where some of the answers to the revision activities in this unit will be put on the poster. You can use some words, but you will mainly be using pictures, diagrams and individual words or phrases to share your knowledge and understanding.

Be creative, make sure that what you produce is eye-catching and interesting as well as scientifically correct.

You will need:
- coloured pens or pencils
- different craft paper
- glue
- large sheet of paper
- pictures from magazines

On your revision rich picture poster, draw pictures or diagrams which will explain what each of these words mean.

(anticlockwise) (axis) (clockwise) (equator) (gas giants) (Moon)

(orbit) (rocky planets) (rotate) (solar system) (star) (Sun)

Revision approach

Mnemonic

A mnemonic is something that helps you learn and memorise a lot of information as something smaller so that it can be stored in your long-term memory. For example, to remember all the colours of the rainbow is a lot of information, so to make it easier, use the first letter of each colour and make a fun sentence that is easy to remember. Then if you are asked to name the colours, the words in the sentence will help you remember all of them.

Ripen	**R**ed
Off	**O**range
Your	**Y**ellow
Green	**G**reen
Bananas	**B**lue
In	**I**ndigo
Vinegar	**V**iolet

Activity 2

Write your mnemonic on your rich picture poster to show the order of the planets in our solar system starting from the nearest planet to the Sun. You might already have one that you have made and used before, if not make a new one that will help you remember the order of the planets from the Sun.

Check first that you know the correct order of the planets before you put your mnemonic on your rich picture poster.

Activity 3

1   Copy the table, on the next page, onto your rich picture poster. Research the information needed to complete the table and choose an interesting fact for each planet.

2   Use your data to answer these questions.

   a   Which is the largest planet? How do you know?

   _____

   _____

   b   How far away from the Sun is Saturn?

   _____

   c   Which planets are the closest to each other?

   _____

   d   What pattern do you notice about the orbit of the planets around the Sun?

   _____

   e   Pluto is not on the table, what happened to Pluto in 2006?

   _____

	Sun	Mercury	Venus	Earth	Mars	Jupiter	Saturn	Uranus	Neptune
Distance from the Sun									
Time taken to orbit the Sun									
Diameter									
Rocky or gas planet									
Interesting fact									

Activity 4

Look at the picture below of Stage 6 learners using a flashlight and globe to model and explain day and night.

You will need:
- globe
- flashlight

The teacher gave the learners these scientific words to use in their explanation:

flashlight	axis	spin	day	night	24 hours
lit	rotation	shadow	tilted	globe	

Here is their explanation.

You need a flashlight and a globe, and a dark place to work. Hold the globe and shine the flashlight on it. Where the light is on the globe it is day, where the light is not on the globe it is night.

1   Which words did they miss out? _____

The teacher read their work and said that they had missed out some important information and that they needed to edit their work and improve their explanation about how day and night occurs.

2   What information did they miss out? How could their work be improved?

_____

_____

3 Rewrite the explanation, making sure that your version is an improvement and that you use all the scientific words that the teacher listed.

_____

_____

_____

_____

Add the explanation you wrote to your rich picture poster.

The Moon

Do you remember?

The Moon

The Moon is not a planet, it is a satellite, it orbits the Earth and is the Earth's only **natural satellite**. Scientists used to think the Moon was about 4.5 billion years old, but because they have a sample of rock brought back by the first people to walk on the Moon, they now think it is slightly younger.

The Moon does not have an atmosphere, it has no wind and no weather but there is water on the Moon.

The Moon is 402 336 km from the Earth and takes approximately 27 days to orbit the Earth, or to be exact, 27 days, 7 hours and 43 minutes. This is called a **lunar month**.

As the Moon orbits the Earth, it spins very slowly on its axis. It takes the Moon one full orbit of the Earth to turn on its axis once. This is why the same side of the Moon always faces the Earth, it does not matter where on Earth you look at the Moon you will always see the same side.

So, the time it takes for the Moon to spin on its axis is almost exactly the same as the time it takes to orbit the Earth which means the Moon always keeps the same side pointing towards the Earth.

Activity 1

Draw five Moons on your rich picture poster and place one answer to each of these questions in each Moon. Use the revision information and research to help you answer the questions.

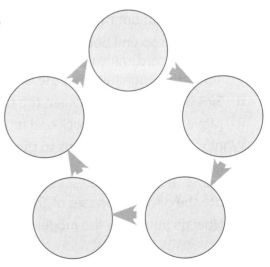

1 Who was the first person to step onto the Moon?
2 Complete this sentence using the correct word.
   The Moon is the Earth's ……………..
   Which of these words is correct? Why?
   Which words are not correct?
   Explain what they are.
   star   satellite   solar system
3 How long does it take for the Moon to orbit the Earth?
4 Explain why we can see the Moon from the Earth?
5 Which statement is correct?
   a People on different parts of the Earth see a different side of the Moon.
   b People on different parts of the Earth see the same side of the Moon.

Do you remember?

The phases of the Moon are the different ways the Moon looks from Earth over a lunar month. As the Moon orbits the Earth, the half of the Moon that faces the Sun is lit up. The phases of the Moon are the shapes of the part of the Moon that is lit by the Sun that we can see from Earth.

Waxing crescent Moon

Activity 2

Make the phases of the Moon

You will need:
- white card: 5" × 3.5" (12.7 cm × 8.8 cm)
- blue card: 5" × 4.5" (12.7 cm × 10.2 cm)
- black card: 7" × 3.5" (17.7 cm × 8.5 cm)
- a circular object to draw around that will fit inside the blue card and leave a space around the edges
- scissors
- glue

1 You are going to make a model to show the phases of the Moon. You might need to revisit a book or the internet to remind yourself of the different phases. As you read these instructions look at the pictures.

   a Measure and cut to size the blue, black and white card.

   b On the blue card, draw around your circular object and carefully cut out the circle.

   c Glue the top and bottom edges of the white card onto the blue card, making sure that the black card can be slotted through between the white and blue card.

   d Slot the black card between the blue and white card.

   e Move the black card back and forth.

2 What do you notice happens to the Moon (the white circle)?

_____

3 Create the different phases of the Moon by moving the black card in and out of the slot.

   a How do you make the model show the Moon waxing?

_____

   b How do you make the model show the Moon waning?

_____

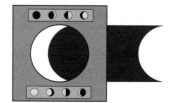

When you have practised making the Moon wax and wane and shown the different phases of the Moon, draw the different phases and write their names on the blue card, so that anyone using your model will know which phase they make.

Activity 3

The phases of the Moon model

Think about the model you have just made.

1   How well do you think the model worked in helping to explain the phases of the Moon?

_____

_____

2   What part of the phases of the Moon did the model not help to explain?

_____

_____

Science in context

The future of the Moon

The first astronauts walked on the Moon on July 20th 1969, when Apollo 11 landed on the Moon. They placed scientific instruments on the surface of the Moon and took samples of Moon rock which they brought back to Earth.

Over the years, scientists have studied the Moon and decided that there is evidence of water on the Moon. They also have evidence of iron and titanium in lunar rocks. On Earth, iron is used to make steel, which is used in many different ways, for example, in buildings, medical instruments, ships, trains, cars, and electrical appliances such as washing machines. Titanium is used for prosthetics (artificial body parts such as arms, hands and legs), tennis rackets, scissors, surgical tools and mobile phones.

Governments of different countries and private companies around the world are planning to send people to the Moon and in the future, there might be research centres where scientists will work. There might even be hotels on the Moon if space tourism (tourists taking trips into space) can get people safely to and from the Moon. There are also plans to use the Moon as a place where scientists can study the effects of living on the Moon before they send humans to live on Mars.

Activity 1

1   Imagine the future when humans can safely travel to and from the Moon. Think about the impact that this kind of space travel will have on the Moon.
    - How do you think the Moon will change?
    - What problems do you think human tourists and scientists will create on the Moon?
    - What do you think will be the good things that could happen as a result of space tourism to the Moon and scientists working on the Moon?

2   Talk to a partner or someone at home about humans travelling to and living on the Moon. Discuss what would be the positive things about doing this, and what the problems might be. On your rich picture poster, list three positive things and three problems that might result from humans travelling to and from the Moon.

Activity 2

Go back to your rich picture poster, read through and check it.

1   Is there anything that you would like to add or change?

2   Think about what you have remembered from previous learning, what do you understand better now? What has helped you develop your understanding?

    _____

    _____

3   If there is anything that you are unsure of, write it on a sticky note and put it on your poster.

4   What new knowledge have you learnt while you have created your rich picture poster?

    _____

    _____

5   When you are happy with your poster display it in your classroom or a wall in your home and share it with someone else.

Revision quiz

Revisit the work that you have done in this unit, check your rich picture poster, read through and remember the scientific words and information. Use your rich picture poster to remind you about Earth and Space. When you are ready complete this short quiz.

As you work through it you can help yourself by:

- Reading each question carefully – check you understand the question.
- Look for key words, use them in your answer.
- Answer the question in your mind first before you write it.
- If the question is multiple choice decide which answers you know are definitely incorrect, then think about the answers that are left, to decide which one is correct.
- Check your answer to make sure that you do not want to make any changes.

Answer the questions on a separate sheet of paper or in your notebook.

1   Which of the following are not planets in our solar system?

    Jupiter   Earth   Moon   Mars   Sun   Saturn   Pluto   Uranus

2   A Stage 6 learner made this mnemonic to remember the order of the planets from the Sun.

    Many Vets Have Mice Just So They Nibble Plants

    Is this mnemonic correct?   Yes No

3   How long does it take for the Earth to orbit the Sun?

   a 45 days

   b 225 days

   c 365 days

4   True or false?

   a Pluto was downgraded to a dwarf planet.

   b The Sun is a planet.

   c The Sun orbits the Earth.

   d The Earth rotates on its axis.

   e Everywhere on Earth it is always daytime.

   f The Earth orbits the Sun, and the Moon orbits the Earth.

   g One complete rotation of the Earth on its axis is called a day.

5   Write definitions for each of the following words:

   a Lunar month

   b Phases of the Moon

   c The solar system

6   Name three different phases of the Moon.

7   Some Stage 6 learners have written these sentences. Check the sentences to see if they are correct or incorrect. If they are incorrect, change the underlined word so that the sentence is correct.

   a When the Moon is viewed from the Earth it appears to change <u>colour</u>.

   b These apparent changes in the shape of the Moon are called <u>phases</u>.

   c We can only see the <u>unlit</u> side of the Moon.

   d At different points in the Moon's orbit around the <u>Sun</u>, different amounts of the Moon's sunlit side <u>face</u> the Earth.

8 Three of the phases of Moon are missing, what are they?

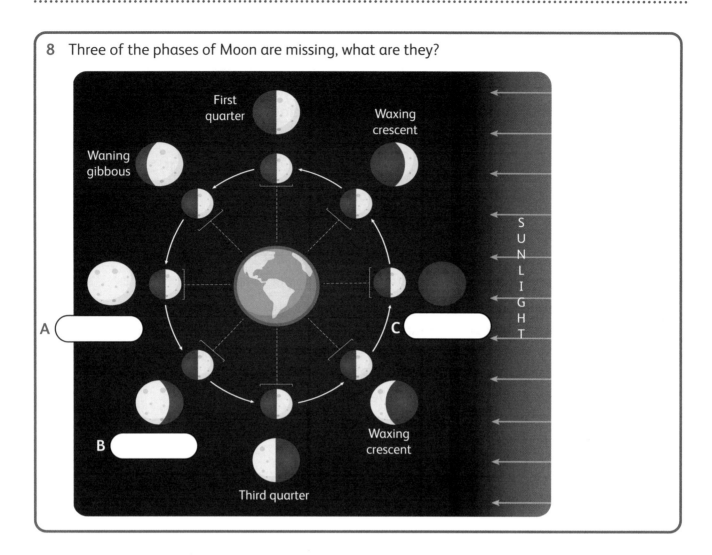

Revision Test

Answer the questions on a separate sheet of paper or in your notebook.

1 The human body has developed defence mechanisms, ways the body uses to help fight infectious diseases. Complete this table. [6]

How microbes can enter the body	Body's defence mechanisms
Openings such as the mouth and nose allow microbes into the stomach	
Different body surfaces such as the eyes, ears	
Breaks, cuts or insect bites on the skin	

2 Name three irreversible changes. [3]

3 Some Stage 6 learners have written the sentences below. Check the sentences to see if they are correct or incorrect. If they are incorrect, change the underlined word so that the sentence is correct. [4]
   a When the Moon is viewed from the Earth it appears to change shape.
   b These apparent changes in the shape of the Moon are called tides.
   c We can only see the unlit side of the Moon.
   d At different points in the Moon's orbit around the Sun, different amounts of the Moon's sunlit side face the Earth

4 Here are five sentences explaining how fossils are formed. In each sentence one word is incorrect. Which word is incorrect and what should the word be? [5]
   a Fossils are the living remains of parts of plants and animals from millions of years ago.
   b The plant or animal was covered after death by mess such as mud and sand.
   c In time, the sediments softened to form rock and the bodies inside the sediments formed fossils, called cast fossils.
   d Some fossils are not formed by the actual plant or animal body but by things left behind by the animals or plants, such as eggs, solid wastes (their blood) and tracks.
   e These fossils are known as track fossils.

5 a Match the labels to the numbers. [5]

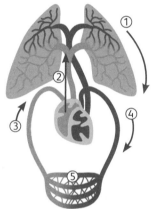

oxygen rich blood, rest of the body, oxygen low blood, lungs, heart
   b Which carries blood back to the heart and lungs, veins, or arteries? [1]
   c What are tubes that carry blood away from the heart called? [1]

6 a Describe the difference between a food chain and a food web. Use these words.

food web   habitat   feeding relationships (links)   living things   energy transfer [4]
   b There is an error in this food chain. What is the error and why? What should the answer be? [2]

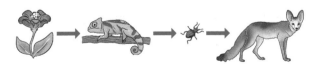

7   When a log of wood burns name: [3]

  a   One product   b   The reactant

  c   Is this a physical or a chemical change?

8   Describe what happens when you inhale and exhale.   [2]

9   a   Name one change that only occurs during puberty in girls.   [1]

  b   Name one change that only occurs in boys.   [1]

  c   Name one change that occurs in both. [1]

10 Describe the relationship (link) between the angle of incidence and the angle of reflection.   [4]

11 Use the words to complete the sentences below.   [5]

  sexual reproduction   sperm reproduce   eggs

  Humans need a male and a female to _____. The male produces _____ and the female produces _____. To create a new offspring (baby), the sperm must join with an egg, this is called _____.

12 In the respiratory system, what happens to the thorax when a person exhales (breathes out)?   [2]

13 Name two materials that are thermal insulators and one material that is a thermal conductor.   [3]

14 True or false?   [4]

  a   In a series circuit, all the components are connected in the same loop.

  b   A parallel circuit must have more than one cell to work.

  c   If components are connected in more than one loop, this is called a parallel circuit.

  d   In a series circuit adding more bulbs results in the bulbs becoming brighter.

15 Explain the term 'bioaccumulation'.   [5]

16 Match the word to the sentence.   [5]

  A   suspension

  B   insoluble

  C   solution

  D   soluble

  E   solvent

    a   A liquid used to dissolve a solid.

    b   When a substance has dissolved in a liquid, the liquid is called a _____

    c   A substance that dissolves in water.

    d   A substance that does not dissolve in water are called _____

    e   When a solid is mixed with a liquid but does not dissolve in it, the liquid becomes cloudy, this is called a

      _____

17 Give an example of balanced forces when an object is   [2]

  a   Stationary

  b   Moving

18 Draw a force diagram for this picture.   [4]

19 Which is the correct answer?   [1]

  a   Gases sometimes have mass.

  b   Gases always have mass.

  c   Gases do not have mass.

20 If a pencil is placed in a glass of water, how will it appear to the person looking at it.   [1]

  a   Normal

  b   Upside down

  c   Bent

  d   Invisible

Total: 75 marks